# Person-Centered
Approaches for Counselors

# THEORIES FOR COUNSELORS SERIES

Person-Centered Approaches
for Counselors

Jeffrey H. D. Cornelius-White
*Missouri State University*

# Person-Centered
## Approaches for Counselors

### Jeffrey H. D. Cornelius-White
*Missouri State University*

Los Angeles | London | New Delhi
Singapore | Washington DC | Boston

Los Angeles | London | New Delhi
Singapore | Washington DC | Boston

FOR INFORMATION:

SAGE Publications, Inc.
2455 Teller Road
Thousand Oaks, California 91320
E-mail: order@sagepub.com

SAGE Publications Ltd.
1 Oliver's Yard
55 City Road
London, EC1Y 1SP
United Kingdom

SAGE Publications India Pvt. Ltd.
B 1/I 1 Mohan Cooperative Industrial Area
Mathura Road, New Delhi 110 044
India

SAGE Publications Asia-Pacific Pte. Ltd.
3 Church Street
#10–04 Samsung Hub
Singapore 049483

Acquisitions Editor: Kassie Graves
Editorial Assistant: Carrie Montoya
Production Editor: Tracy Buyan
Copy Editor: Deanna Noga
Typesetter: Hurix Systems Pvt. Ltd.
Proofreader: Theresa Kay
Indexer: Maria Sosnowski
Cover Designer: Anupama Krishnan
Marketing Manager: Shari Countryman

Copyright © 2016 by SAGE Publications, Inc.

All rights reserved. No part of this book may be reproduced or utilized in any form or by any means, electronic or mechanical, including photocopying, recording, or by any information storage and retrieval system, without permission in writing from the publisher.

Printed in the United States of America

*Library of Congress Cataloging-in-Publication Data*

Cornelius-White, Jeffrey H. D.

Person-centered approaches for counselors / Jeffrey H.D. Cornelius-White, Missouri State University.

pages cm. — (Theories for counselors series)

Includes bibliographical references and index.

ISBN 978-1-4522-7772-1 (pbk. : alk. paper)

1. Counseling—Study and teaching. 2. Client-centered psychotherapy. I. Title.

BF636.65.C67 2015

158'.9—dc23        2014042366

This book is printed on acid-free paper.

15 16 17 18 19 10 9 8 7 6 5 4 3 2 1

# Brief Contents

| | |
|---|---|
| Series Preface | ix |
| Acknowledgments | xiv |
| Introduction | xvii |
| Chapter 1: The Therapist in Relationship | 1 |
| Chapter 2: Client | 11 |
| Chapter 3: Evolution of the Person-Centered Approach | 23 |
| Chapter 4: Multiculturalism | 39 |
| Chapter 5: A Case Illustration of Person-Centered Counseling | 53 |
| Chapter 6: Conclusion | 67 |
| References | 73 |
| Index | 81 |
| About the Author | 85 |

# Detailed Contents

| | |
|---|---|
| Series Preface | ix |
| Acknowledgments | xiv |
| Introduction | xvii |
| **Chapter 1: The Therapist in Relationship** | **1** |
| Therapist and Relationship Factors | 1 |
| The Necessary and Sufficient Conditions | 2 |
| Summary | 9 |
| **Chapter 2: Client** | **11** |
| Contact, Incongruence, and Perception | 13 |
| Client-Directed, Outcome Informed, Client Hero | 17 |
| Actualizing Tendency | 18 |
| Nondirective Attitude | 19 |
| Conclusion | 20 |
| Summary | 21 |
| **Chapter 3: Evolution of the Person-Centered Approach** | **23** |
| Classical Client-Centered or Nondirective Therapy | 24 |
| Focusing | 27 |
| Intersubjectivity | 28 |
| Experiential, Emotion-Focused | 30 |
| Interdisciplinary Research and Practice | 33 |
| The Futurism of the Person-Centered Approach and Its Implications | 36 |
| Summary | 37 |

Chapter 4: Multiculturalism                                              39
   Central Concepts in Multicultural Counseling Competence              40
   Understanding Client's Worldview: Cultural Elements to Empathy       42
   Culturally Appropriate Interventions and Unconditional
      Positive Regard                                                   43
   Social Justice Advocacy Competencies                                 45
   History of Multiculturalism and the Person-Centered Approach         46
   Ethical and Religious Concepts and the
      Person-Centered Approach                                          47
   The Paradox: A Universal System of Adaptability to Differences       50
   Summary                                                              51

Chapter 5: A Case Illustration of Person-Centered Counseling            53
   Deena                                                                53
   Casey                                                                53
   Session 1: First Encounter                                           54
   Session 2: Scared, Really Scared                                     56
   Session 3: Some Relief                                               58
   Session 4: A Possible Misunderstanding                               58
   Session 5: Nonstop Talker                                            60
   Session 6: I Miss My Mom, Bad. She Was a Bitch to Deal With          61
   Session 7: Talkative Again                                           63
   Sessions 8 and 9: No Show                                            64
   Questions for Your Consideration                                     64
   Summary                                                              65

Chapter 6: Conclusion                                                   67
   PCA Is Love                                                          67
   Courageous, Compassionate Confrontation                              68
   Perception                                                           68
   Linking the Relational and the Client Conditions                     68
   It Figures                                                           69
   Understand, Verify You Understand, and Improve                       70
   Power Within, Power With, Power Over                                 70
   The Ever-Reaching Actualization Process: Give and Receive            71
   Summary                                                              71

References                                                              73
Index                                                                   81
About the Author                                                        85

# Series Preface

"Theories for Counselors" provides practical applications of major theories from a common factors, multicultural perspective. What does that mean? Let's break it down.

The authors in the "Theories for Counselors" series are highly experienced counselors with extensive knowledge and expertise concerning the theory that they present. They present each theory from an applied perspective, asking themselves, "How is this concept useful in actual clinical practice?" It may surprise you to know this, but Freud's work can be (and is) applied day in and day out in modern counseling. (If this surprises you, it could indicate that you have not been taught Freud well.) He believed that the relationship between the client and clinician was of utmost importance; he believed that his patients needed to feel comfortable speaking their mind; he believed that clinicians needed to listen with attentiveness and tact. Freud's legacy, as is shown in the first book of this series, *Psychoanalytic Approaches for Counselors*, has been revised and revisited, but its therapeutic usefulness remains, and for each theory that is presented, therapeutic utility is utmost on the minds of the authors as they present material to their readers.

Each book begins by addressing the two most vital themes common to any counseling theory: the client and the therapeutic relationship. Why have we picked the client and the therapeutic relationship as the two most important themes? The reason is called the *common factors hypothesis*, and this is where research comes in. The common factor hypothesis is the result of decades of research that has compared various schools of counseling and psychotherapy. Contrary to prior belief, it has been convincingly demonstrated that research in general finds no significant difference in how effective the various therapies are. These findings, predicted by Rosenzweig (1936) nearly 80 years ago, began to be empirically demonstrated in the mid-1970s (Luborsky, Singer, & Luborsky, 1975; Smith & Glass, 1977). Research confirming the relative equivalence of bona fide therapies has accumulated since that time (e.g., Ahn & Wampold, 2001; Lambert, 1992; Lambert & Barley, 2001; Lambert & Ogles, 2004; Wampold et al., 1997).

What does this mean? It means that instead of therapeutic improvement being due to specific ingredients prescribed by different theoretical schools of counseling and psychotherapy, positive therapeutic change can be attributed to factors that are common to all bona fide therapies. Additionally, these factors can be broken down into four categories: client variables (40% of change), relationship variables (30%), hope and expectancy (15%), and theory or technique (15%) (Duncan, 2002; Lambert, 1992) (see Figure 1).

As we see, the client and the relationship accounts for the vast majority of therapeutic change, and as such, should be centrally located in the presentation of any counseling theory.

Interestingly, the history of counseling begins right where research predicts: in an intense relationship between one person who wants help and another person wanting to help. Sigmund Freud, who inaugurated themes that continue to organize the counseling profession, described and redescribed the origins of psychoanalysis. Two major components in his descriptions were the famous first patient of psychoanalysis, Bertha Pappenheim (referred to in case studies as "Anna O."), and the relationship she had with her doctor, Josef Breuer, Freud's friend and colleague at the time. Though Freud revised his opinions on many things about that famous case (as he did about almost everything), what remained constant was the fact that he saw something of primary importance in that case—the "talking cure" that occurs between a patient/client and doctor/counselor.

Thus the origins of counseling display a deep consonance with the latest in empirical research, and it is this consonance that is the underlying theme

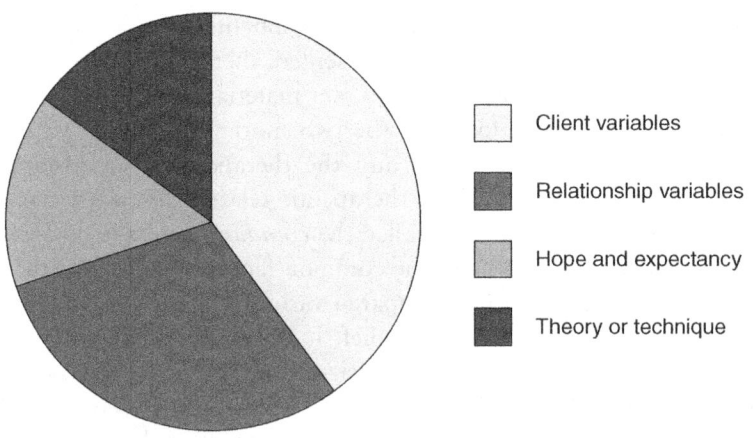

**Figure 1**  Common Factors

*Source*: Lambert, M. J. (1992). Psychotherapy outcome research: Implications for integrative and eclectic therapists. In J. C. Norcross & M. R. Goldfried (Eds.), *Handbook of psychotherapy and behavior change* (4th ed., pp. 143–189). New York, NY: John Wiley.

behind the series "Theories for Counselors." Starting with Freud and moving through past and contemporary counseling theories and theorists, the focus remains on the client and the therapeutic relationship, and how this relationship fosters and enhances the client's natural resilience and hope for change. The theory's techniques and the theory itself are important only inasmuch as they provide a common roadmap—a way for both client and counselor to think about where a client has been and where he or she wants to go.

Just as it is important to know that Freud remains useful for contemporary counselors, so it is important to know that Freud began his work against a backdrop of rising racial hatred in Austria and Western Europe, and that while he successfully fled to England in 1938, his sisters perished in the Nazi concentration camps during the Holocaust (Gay, 2006). Thus the counseling enterprise began at a time of extreme racial hatred, which is a sobering and important fact to reflect on; from the inception of counseling in Western Europe and throughout its development worldwide, multicultural awareness and respect for diversity are no mere add-ons but are integral components for the practice of counseling. In addition, another important group membership—gender—has assumed greater and greater importance in the counseling field; its central importance imbued the case of Bertha Pappenheim, which has been deemed the founding one for psychoanalysis and hence for all that followed.

Counselors must practice from a culturally aware place rather than one that would seek to downplay the impact of race, gender, and other important group affiliations on our clients' lives. Sue, Arredondo, and McDavis (1992) provided a conceptual framework for organizing the types of competencies needed by a culturally skilled counselor, saying that he or she becomes aware of his or her own assumptions, actively attempts to understand differing worldviews, and actively develops culturally sensitive intervention strategies and skills. Sue (2001) expanded his conceptual framework into a multidimensional model of cultural competence; this model was primarily focused on racial and ethnic minority groups, though he did also recognize that it might be applicable for other groups including those of "gender, sexual orientation, and ability/disability" (p. 816). Such topics are now recognized as rightfully fitting within the context of multicultural counseling (Conyers, 2002; Pope, 2002; Richardson & Jacob, 2002). In addition, Smith and Richards (2002) point out the obligation that counselors have to address issues of religion and spirituality as multicultural issues.

D'Andrea and Daniels (2001) provided a multicultural framework for working with clients that is RESPECTFUL and inclusive of Religion and spirituality, Economic class, Sexual identity, Psychological development, Ethnic/racial identity, Chronology, Trauma, Family, Unique physical abilities and disabilities, and Language and location of residence. Similarly, Hays

(2008) outlined a model that emphasized nine cultural influences in relation to specific minority groups that counselors should be ADDRESSING: Age, Developmental and other Disabilities, Religion, Ethnicity/race, Social status, Sexual orientation, Indigenous heritage, and Gender. These models help counselors deliver diversity competent services and pay attention to all the potential resources that a client brings to the counseling encounter. Ultimately, respect for diversity and celebration of all aspects of culture and group membership should lead to a more nuanced understanding of the client and the sometimes-hidden strengths that he or she possesses. Better knowing a client enhances the richly relational counseling encounter that began with Freud's work.

Once again, this is consonant with the common factors approach. The common factors approach can be applied to, and makes sense of, any counseling theory, beginning with Freud and psychoanalysis. According to this approach, all bona fide counseling theories do the same thing, though they describe it using differing terminology. One analogy is traveling to a particular destination, say New York City. There is no one right way to get there—it depends on where you are starting from, whether you want to fly, drive, or take the train, and whether you want to get there by a direct route or take a scenic one. Each unique route is analogous to a different counseling theory. The destination is the same—in a travel scenario, getting to New York City, and in a counseling scenario, achieving positive treatment outcome.

In their book, *The Heroic Client*, Duncan and Miller (2000) put it this way—they seek to "(1) enhance those factors across theories that account for successful outcome; (2) encourage the client's unique integration of different theories; and (3) selectively apply diverse ideas and techniques as they are seen as relevant by the client" (p. 146). Miller has talked about the need for clinicians to know different theories because they serve as language resources to connect with the client. In this view, theory is a way to connect with clients; if one language that I'm using—for instance, solution focus therapy—doesn't appear to be the language that the client is speaking, then I should use other theoretical languages that might allow me to communicate better with my client. The test of theory is in how well it accords with each individual client's culturally influenced worldview and how useful it proves to be in the context of the therapeutic encounter.

"Theories for Counselors" will help you consider theories from the perspective of the client and what makes sense to her or him. It will show that theory and technique are good inasmuch as they aid clients in understanding their present situation and what they need to do to improve it. Finally, the series will help you situate the work of counseling within a sociocultural framework that takes into account client uniqueness, universality, and important group affiliations to enhance and activate client resources.

## A BRIEF PRIMER ON THE COMMON FACTORS: HYPOTHESIS

- Suspecting that there were characteristics that all effective therapies (and therapists) shared, Rosenzweig (1936) makes the claim that there are common factors upon which good counseling rests, regardless of theoretical orientation.
- Researching further, scholars use meta-analysis—a summary about relevant empirical studies or an "analysis of analyses"—as a tool to compare treatment outcome studies.
- Gaining consensus from these meta-analytic studies, researchers suggest that positive therapeutic treatment outcome is due to the following four groups of factors:
  1. Client/extratherapeutic factors: 40%
  2. Therapeutic relationship factors: 30%
  3. Placebo/expectancy for change/hope: 15%
  4. Theory/technique: 15%
  (Duncan, 2002)
- Correcting the above breakdown, later researchers suggest that while the common factors hypothesis has been supported—"these shared, curative factors drive the engine of therapy" (Hubble, Duncan, Miller, & Wampold, 2010, p. 28)—it has been somewhat oversimplified. The factors aren't discrete but are complexly interrelated.
- Focusing on effective therapists rather than effective therapies should be emphasized, since there are large differences in effectiveness among clinicians, and "Using formal client feedback to inform, guide, and evaluate treatment is the strongest recommendation" that came from a comprehensive update of common factors (Hubble et al., 2010, p. 424).
- Condensing this into a sentence, it means that from a common factors perspective, counselors use a theoretical roadmap—i.e., employing language that makes sense to the client—that helps activate dormant client resources and enhance client hope and expectancy for change, within the context of a therapeutic relationship that "is characterized by trust, warmth, understanding, acceptance, kindness, and human wisdom" (Lambert & Ogles, 2004, p. 181) and that can allow for modifications and changes based on feedback from the client.

Finally, I direct the reader to the companion website for this book and series, http://study.sagepub.com/theoriesforcounselors. There you will find extended discussions of topics that are mentioned briefly in the text, topics that are not addressed in the text but that might be useful to know when studying for comprehensive or licensing exams, definitions of terms, and supplemental exercises and activities. In general, if a topic is not covered or is covered in detail in the printed text, please search the website, as it will in all likelihood be discussed there.

# Acknowledgments

I acknowledge the help of Michelle Ciesielski and Chris Carver who helped hold my voice, both in this book and in general during a difficult time. I also acknowledge the many clients, students, colleagues, mentors, family, and friends who together honed my ability to learn, research, integrate, and practice the person-centered approach. I am forever in your debt.

*This work is dedicated to four generations of women who have inspired me to be who I am by being who they are: Nana, Mom, and the two AJs. I am grateful for the time I have had with them and the permanent etch these relationships have had on my heart. Avery Jane, you are awesome. I have never experienced such a frequent sense of amazement in the world as I have with you.*

*It is also dedicated to four generations of men who have inspired me to endure with surprising grace despite untold suffering: Grandpa Chuck and Slim, Dad, Herbie and Chebby, and Evan. Evan Riley, you will never know how much I love you. Your humor and resilience shine brilliantly.*

# Introduction

Perhaps more than any other major theorist, Carl Rogers was an advocate for *common factors*. He put the two central themes of the "Theories for Counselors" series—the client and the therapeutic relationship—as the two foundational blocks of not only person-centered therapy, but also his view of all psychotherapy. Rogers saw his clients as the agents of change in the therapeutic encounter. He saw the relational conditions as the background that encouraged clients' success. In his seminal paper, "The Necessary and Sufficient Conditions," he laid the groundwork for common factors research and practice for decades to come (Rogers, 1957). He asserted that *how* a counselor does something is more important than *what* a counselor does. Specifically, it is the counselor's *attitude toward* and *adjustment to* how the client receives the counselor's actions that is crucial.

Carl Rogers led a long, highly engaged and productive life. He was born in 1902 and died in 1987, a descendant of the first Americans on both his maternal and paternal sides dating back to the 1630s, showing deep roots for the American spirit of self-actualization and democratic communication present throughout his works. As a child near Chicago, Illinois, he was inquisitive and hardworking, enjoying the outdoors and reading and completing many chores. His mother was particularly religious, and the family spent considerable time in Bible study and prayer.

He went to the University of Wisconsin to study agriculture. At 20 years old, he was selected to represent the YMCA at the World Student Christian Federation conference in China. He traveled to China, Japan, Korea, the Philippines, and Hawaii, and interacted with some of the world's political and religious leaders, including the ex-chancellor of Germany, the mayor of Seoul, Nobel Peace Laureate John S. Mott, and dozens of other historically important figures. During this trip, he kept a diary that foreshadowed many of his contributions to psychology and revealed his personality in an intimate way (Rogers & Cornelius-White, 2013). Much of Rogers's work in writing and in person shows this candor, warmth, and adventurousness, all with an understated vigor. After his return, he switched his course of study to theology and moved to New York City. In graduate school, he studied psychology and then moved to Rochester, New York, where he wrote his first published books.

He was offered a full professorship at the University of Ohio and then took such positions at the University of Chicago and the University of Wisconsin. During these positions, Rogers put forth many of the theoretical contributions and research studies that became the basis for person-centered therapy, including *Counseling and Psychotherapy* (1942) and *Client-Centered Therapy* (1951) and *On Becoming a Person* (1961). He also inspired hundreds of works on learner-centered education, a research tradition nearly as rich as that of person-centered therapy (Cornelius-White, 2007c). After some conflict and disappointment at the University of Wisconsin, in 1964, he moved to La Jolla, California, where he held research and service positions for the rest of his life, but did not return formally to academia. Rogers became particularly active in encounter groups (see the Academy Award winning film *Journey Into Self* [1968]) and peace work, traveling much of the time for professional development workshops and to conflict hot spots such as Northern Ireland, South Africa, and the USSR. From age 20 to 85, Carl's writings show his thorough engagement in multicultural communication. The *Carl Rogers Reader* edited by Kirschenbaum and Henderson (1989) provides an accessible entry point to many of these personal and professional writings. Kirschenbaum's (2008) biography is the definitive work on his life, based on unique access to Carl directly while alive and in documents released after his death.

In repeated surveys across the last three decades, Carl Rogers has been voted the most influential American psychotherapist by counseling practitioners (Cook, Biyanova, & Coyne, 2009). Despite widespread appreciation for Rogers, many mental health researchers and practitioners alike tend not to take him seriously. Many have heard the joke about the suicidal client. The person-centered counselor continued to reflect the client's statements like a parrot even as the client prepared to and then did jump out a window. Such shallow perceptions of the person-centered approach are stunningly all too common. In some ways, Rogers has been a victim of his own success. Current researchers less frequently cite or acknowledge his work, and practitioners tend not to recognize the specific contributions he made. In many ways, the person-centered approach to counseling has become so interwoven with basic or fundamental skills, that the paradoxical depth, challenging discipline, and revolutionary qualities of the person-centered approach can be lost.

*Person-Centered Approaches for Counselors* will help counselors-in-training, counselors, and counselor researchers alike to appreciate the unique contributions—both classical and ongoing—that person-centered therapy offers. The person-centered approach is the phoenix of common factors, rising from the ashes of neglect again and again to provide fresh vitality for advancing both counselors-in-training and the field of counseling in general (Cornelius-White, 2002). What are some of those unique contributions?

Chapter 1, "The Therapist in Relationship," reveals how *who you are matters more than what you do*. The person-centered approach is first and foremost a way of being.[1] It's a stance, a set of attitudes. Chapter 1 emphasizes trust in persons, both oneself as a counselor and the resilience and capacity of one's clients. The "If-Then" hypothesis of Rogers's (1957, 1959) "Necessary and Sufficient Conditions" predicts that IF you can *genuinely* provide and your clients can perceive *empathy* and *unconditional positive regard*, THEN you can trust that your clients will get better. Abundant research from within psychotherapy, education, parenting, and other domains has shown how in humble, loving relationships, people will suffer less, accept themselves more, find solutions, and make changes in their perspectives that lead to richer, more satisfying lives (Cornelius-White, 2007c; Cornelius-White & Motschnig, 2010; Norcross, 2002). Chapter 1 provides examples of each of these core relational conditions in practice, including excerpts from therapy transcripts and challenging scenarios for you to consider. It will help counselors understand the indisputable value of deep and nonpossessive loving relationships with their clients. After all, is there anything in life that you appreciate more than a good friend, family member, or true love?

Chapter 2, "Client," helps counselors understand how *persons have the potential to grow and develop* in both horrific and ideal circumstances. Rogers's *actualization tendency* provides the theoretical basis for this motivation, capacity, and resilience. In many ways, the person-centered approach is about *confronting experience*. It's about *courage* on the part of both you and your clients to *experience* and *accept* suffering, destruction, rebirth, happiness, and renewed suffering. The person-centered approach is learning how to be with people in their *poignancy*, bringing to life and focus the challenges they experience. Practicing the person-centered approach involves a strong paradox: The more you can accept how helpless you or your client is to avoid suffering, the more power you and your client can discover. Beginning counselors often struggle with wanting to "fix" clients immediately rather than learn to "be with" clients. They often view

---

[1] *A Way of Being* is the title to Carl Rogers's (1980) influential collections of personal and philosophical essays in the final decade of his life.

power as "power over" another rather than "power with," thus robbing clients of the gift of not only *having a solution* to their problem but also *solving* their problem. At the same time, they miss emotions or, worse still, believe they are being "supportive," trying to persuade clients they are OK and encouraging them not to experience their emotions and personal meanings. Too frequently, counselors think this is being "empathic," not realizing that the avoidance or flattening of emotion and meaning is precisely the opposite of evocatively empathizing (Martin, 2010). The challenge for a counselor influenced by person-centered approaches is to practice and learn the discipline needed to empathize accurately and accept clients in the moment, and grow oneself to meet this challenge to prevent squandering clients' potential. Chapter 2 highlights evidence from psychotherapy process research to show how the client is a hero, actively molding the relationship and the counselor's contributions into resources for life success. Drawing on examples from Bohart and Tallman (1999) and Duncan, Miller, and Hubble (1999), it will emphasize often-surprising perceptions clients have of their counselors. It will show examples of how they take unexpected actions to transform the relationship in ways that help them beyond the counselor's intent.

Chapter 3, "Evolution of the Person-Centered Approach," reveals how person-centered approaches to counseling continue to grow through improvements in both understanding and practicing the classical model, modifications (especially related to experiential approaches), and interdisciplinary research both within and especially beyond psychotherapy. As early as 1942, Rogers provided the very first therapy transcripts and some of the first quantitative counseling process and outcome research (Kirschenbaum, 2009). Whether through *relational depth* or the *intersubjective* nature of the person from authors such as Mearns, Bozarth, Barrett-Lennard, or Schmid, classical person-centered therapy has had contemporary advocates that continue to enrich our understanding of the fundamental concepts of client and relationship (Sanders, 2004). Modifications have included *experiential, focusing-oriented,* and *emotion-focused therapy* from authors such as Gendlin, Elliott, and Greenberg (Sanders, 2004). These *experiential* researchers have produced some of the strongest empirical research, using innovative mixed-method designs, gold standard randomized control studies, and influential meta-analyses to demonstrating the equivalence, or in select situations, superiority, of person-centered approaches to counseling in contrast to the dominant medical model approaches (e.g., cognitive behavioral therapy and psychopharmacology). Person-centered approaches provide a wealth of transcripts and practical advice of how a counselor can apply other techniques to adapt to client needs in keeping with the *common factors premise* of accordance with the client's perspective.

Finally, while present from its origins, *interdisciplinary research* has had a renaissance of sorts relating various scientific and applied fields to the person-centered approach. Neuro-scientific examples help counselors see how these latest developments spur improved understanding and practice. For instance, your mirror neurons activate when you watch a person eat an ice cream cone or walk across fire. Mirror neurons thus suggest the biological substratum of empathy. Likewise, there are reliable heart rate and skin conductivity concordances in successful therapy relationships that are not visible in successful ones (Cornelius-White, Motschnig-Pitrik, & Lux, 2013a, 2013b).

Chapter 4, "Multiculturalism," helps counselors see how *universal and culturally specific factors interact* in successful ways between people who are culturally different. By fostering a respectful, empathic relationship that adapts to the perceptions of others, the person-centered approach is uniquely helpful in advancing multicultural counseling goals (Cornelius-White, 2003, 2005; Lago, 2012), including the multicultural (Association for Multicultural Counseling and Development, 1992) and social justice advocacy competencies. Research and practice within the person-centered approach has been at the forefront of multicultural issues from its inception through today. Early multicultural integrations include Carl Rogers's trip in 1922 as a 20-year-old (Cornelius-White, 2012) and research on how the approach reduces prejudice and fascism during and after World War II (Cornelius-White & Harbaugh, 2010). Through the encounter group and civil rights movements in the 1960s and 1970s and into his late life in the 1980s negotiating peace in conflict zones like apartheid South Africa, war torn Northern Ireland, and in cold war "opponent" USSR, Rogers was a beacon of *cross-cultural communication*; a staunch opponent of *social injustice*; and an *advocate* for children, women, people of color, laborers, and anyone disenfranchised by the social situations in which they find themselves (Kirschenbaum, 2009). Indeed, Carl Rogers was nominated for the Nobel Peace Prize when he died. Additionally, person-centered approaches are linked with diverse religious practices that can enrich counselors' practice: the Taoist concept of *wu-wei*, Buddhist concept of *non-attachment*, Christian concept of *grace*, and secular concept of *mindfulness*. Finally, global research with over 5,000 participants has shown how minority practitioners have a greater preference for person-centered approaches than other major approaches to counseling, such as psychodynamic, cognitive behavioral, or systemic (Elliott, Orlinsky, Klein, Amer, & Partyka, 2003).

Chapter 5, "A Case Illustration of Person-Centered Counseling," presents the fictitious case of Deena, a suicidal client with Jennie, a first semester intern. The presentation will describe several weeks' worth of sessions including actions, responses, thoughts, and feelings of both persons. It also provides examples of the concepts covered in the other chapters regarding clients, counselors, the counseling relationship, the feedback loop moment-to-moment and more formally with evaluation of client perceptions, multicultural issues, and elements of different person-centered approaches integrated together. The case shows how a talented, attentive, but clearly imperfect and inexperienced counselor-in-training was unequivocally helpful with a difficult client and some of the challenges they encounter along the way.

Chapter 6, "Conclusion," synthesizes the themes of *Person-Centered Approaches for Counselors* and situates them in relation to the common factors themes of the *Theories for Counselors* series. Like any of the theories explored in the series, person-centered approaches are useful to the extent that they help value and foster the heroic qualities of clients, particularly the accordance of the counselor's approach with the client's perspective, and further the usefulness of the therapeutic encounter. This concluding chapter summarizes and synthesizes the therapist or relational conditions of genuineness, empathy, and unconditional positive regard through the theme of non-possessive love. It emphasizes how counselors can help clients courageously *confront and make meaning* from their challenging but potentially rewarding life *incongruent experiences*. It will highlight the central importance of *client perception*—that empathy, unconditional positive regard, and genuineness are successful only to the extent that the client *perceives* them. Thus, the *target* of a person-centered counselor's statements and the *feedback mechanism* for better counselor responses are *client's perceptions to those responses*. It highlights the need to have *discipline* to *be with*, understand, communicate, and allow clients to identify, work through, find, and attempt solutions to their problems. It presents an original figure that links together the core client and therapist conditions to help the reader grasp and apply them. Chapter 5 discusses the use of quick assessments to improve therapist behaviors and judgment (Duncan, 2013) and how person-centered approaches rely on "power within" and "power with" rather than "power over" clients. Finally, it emphasizes the ever-reaching *actualization process* through suffering and connection that applies to counselors-in-training and counselors, as much as to clients.

**SAGE** was founded in 1965 by Sara Miller McCune to support the dissemination of usable knowledge by publishing innovative and high-quality research and teaching content. Today, we publish more than 750 journals, including those of more than 300 learned societies, more than 800 new books per year, and a growing range of library products including archives, data, case studies, reports, conference highlights, and video. SAGE remains majority-owned by our founder, and after Sara's lifetime will become owned by a charitable trust that secures our continued independence.

Los Angeles | London | Washington DC | New Delhi | Singapore | Boston

# 1

# The Therapist in Relationship

Is there anything in life that you appreciate more than a good friend, family member, or true love? For several years I have gathered journals from students in which they write three things they are most grateful for each day for one week. Inevitably, what they report they are most grateful for is their positive relationships. People are grateful for the love they share with their kids, their partner or spouse, their friends, their parents, or their dog. At a basic level, the central theme of person-centered counseling is highly intuitive. You almost just know that it's true without learning any theory, research, or practice deliberately engaged in for years: Caring relationships help people grow and flourish.

But how do you build, maintain, and improve such caring relationships with people you do not know, people who often hate themselves, or people who are difficult for others to like, love, or even just tolerate being around? The person-centered approach is first and foremost a way of being. It is a stance, a set of attitudes. It offers the hypothesis that who and how you are matters as much as what you do or say.

## Therapist and Relationship Factors

For decades, one the most robust findings in the psychotherapy literature is that working alliance and client perceived empathy are strong predictors of therapeutic outcome (Norcross, 2002). If you have a good working relationship and the therapist successfully conveys an experience of empathy, it's very likely that most clients' symptoms and life situations will improve. Person-centered counseling articulated and stressed relationship factors to a unique degree compared to other approaches. What is becoming increasingly clear in the literature in the last decade is that the therapist, not just the approach or relational qualities, is a strong underlying factor in creating the relationship. In fact, Hubble, Duncan, Miller, and Wampold (2010) argue "available evidence documents that the therapist is the most robust predictor of outcome of any factor ever studied" (p. 38). The best counselors have dropout rates 50% less and improvement rates 50% higher than average counselors (Hubble et al., 2010). A large part of why these best therapists have so much more success appears to be their ability to form, maintain, and/or improve therapeutic relationships. Person-centered counseling provides a clear, empirically validated explanation for the qualities of these relationships to foster in beginning and experienced counselors alike.

## The Necessary and Sufficient Conditions

Rogers (1957, 1959) proposed conditions, which IF they were present, THEN therapeutic change would occur. There are six conditions, though Conditions 3, 4, and 5 are sometimes referred to as the three *core conditions*. They are as follows:

1. Two persons are in contact.
2. The first person, whom we shall term the client, is in a state of *incongruence*, being *vulnerable or anxious*.
3. The second person, whom we shall call the therapist, is *congruent in the relationship*.
4. The therapist is experiencing *unconditional positive regard* toward the client.
5. The therapist is experiencing an *empathic* understanding of the client's *internal frame of reference*.
6. The client *perceives*, at least to a minimal degree, Conditions 4 and 5, the *unconditional positive regard* of the therapist for him [sic], and the *empathic* understanding of the therapist. (Rogers, 1959, pp. 238–239)

In this chapter, we discuss those factors that are primarily associated with the counselor: congruence, unconditional positive regard, and empathic understanding. In contrast, those factors that are *between* the therapist and client (contact and communication) or *within* the client (incongruence and perception) are addressed in the next chapter.

### Congruence

"Congruence is perhaps the most difficult concept to understand, facilitate, develop, measure, and agree upon within the person-centered approach" (Cornelius-White, 2013, p. 168). That being said, a lack of congruence is easy to detect by clients. Clients can often tell when we are distracted or upset without us having to say anything direct about it. In fact, one study (Grafanaki, 2001) showed that congruence was communicated 23 times more by our bodies and voice tone than our words.

So what is congruence? Basically, it's being real, genuine. It's not hiding, but being sincerely unconditional and empathic, not just faking it. There are a lot of synonyms that have been used, each with its own shade of meaning. *Integrated* is perhaps the closest synonym, and it's one that Rogers uses directly within the conditions shown above. Being integrated within the relationship means that during the time you are with a client and in connection to the client (not during every moment of your life or in every relationship in your life—Thank God!), you are self-aware and present. It means that you do not have internal conflicts toward the person that you are not aware of and accepting of. Your verbal, nonverbal, and internal messages are consistent or at least self-accepting regarding any inconsistencies. It also means that the other conditions of unconditional positive regard and empathy are integrated in you, as part of your attitude toward a client.

*Transparence* refers to the sense of what you see is what you get. It's the quality of being open and willing to (but not having to) share your experience. Sometimes people use transparence to refer to communications that come from your congruent state. In other words, you might transparently self-disclose part of your experience of empathy, unconditional positive regard, or other relevant personal experiences. *Authenticity* is another term people use to describe the quality of being who you are, accepting feelings or perceptions that might be difficult and interacting sincerely. However, in person-centered therapy, ideally this authenticity is integrated with the other core conditions and does not just refer to saying whatever is on your mind with the belief that because it's really how you feel, that means it will be therapeutic to share (see Table 1.1).

Table 1.1  Dimensions of Congruence

| Dimensions | Consistency Between | Relationship to PC Theory | Synonyms |
| --- | --- | --- | --- |
| 1 | Attitudes | Sincerely Empathic and Unconditional | Genuineness |
| 2 | Self and Experience | UPR Toward Self | Symbolization |
| 3 | Self, Experience, and Communication | Transparent, Spontaneous Self-Disclosure | Authentic |

*Source*: Adapted from Cornelius-White, J. H. D. (2007). Congruence: An integrated five dimension model. *Person-Centered and Experiential Psychotherapies, 6*, 231.

*Note*: UPR = Unconditional Positive Regard.

## Examples

1. You empathically follow your client, accepting what they say. You are absorbed in their reactions, leaning forward, and reaching with your thoughts to communicate accurately with your heart and your words the most challenging edges of what your client is going through. In this case, you are transparent. What you see (absorbed unconditional empathy) is what you get. You are authentically engaged—consistent inside, through your nonverbal messages, and in your words.

2. You are empathically following your client, when you become aware that what they are telling you about what they did to their partner disgusts you. You notice your internal repulsion and fear for the third party, but you accept it. You hold your experience while continuing to remain focused on how the client experienced frustration and lashed out from feeling helpless in how to respond when their partner had not listened to them. You are genuine in your empathy as conveyed through consistency in your nonverbal messages and words and congruent within your own self in knowing and accepting how your own reactions intersect without having to derail the client's process or your own empathic, unconditional way of being.

3. You are facilitating a group with high school students and two girls are in conflict. Susie tells Maria that she felt betrayed when Maria disclosed a secret they shared to a mutual friend; Maria criticizes Susie for being too sensitive, and Susie calls Maria a bitch. You are not only aware how both persons' perspectives make sense, but also how you yourself are saddened by the interaction they are having after the struggles the two of them had previously overcome. You say to them (and the group), starting slowly and quietly, looking at each of them as you address their experience,

> "I am saddened by how the two of you are talking past each other and not listening. Susie, I know Maria was harsh when she criticized you just now, and Maria, I know Susie was just cussing at you, but I'm also aware that you are both hurt—hurt and maybe betrayed after feeling closer during these last few weeks and believing you could trust each other again."

You are congruent between your inner experience, your actions, and your words. You self-disclose your feeling, but do so with empathy and unconditional positive regard integrated within you.

## Unconditional Positive Regard (UPR)

The second core condition is described by such terms as acceptance, nonpossessive warmth, respect, nonjudgmental interaction, and benign neutrality. It refers to an acceptance of the client's person, their feelings, conflicts, perceptions, solutions, and actions. It is not agreement or endorsement, though clients sometimes infer this. In some respects unconditional positive regard is like love, especially agape love, prized by Christian philosophers.

However, unconditional positive regard is not accurately captured by the maxim: love the sinner but hate the sin. It is a full, but nonpossessive, love for the other person, accepting them for who they are and trusting in the potential for who they can become. Table 1.2 describes some of the key points of UPR.

## Examples

It is important to get that unconditional positive regard is both unconditional and positive to get the concept. Many people equate positive statements (e.g., I think you are doing great.), encouragement (e.g., It will get better. You will make it through this.), or reassurance (e.g., That's a normal reaction. It's not as bad as it seems.) with unconditional positive regard. However, from a person-centered standpoint, these are as opposite to unconditional positive regard as negative statements (e.g., That's wrong.) or discouragement (e.g., You won't make it through this.).

What all these statements have in common is that they are conditional. This is where the idea of benign neutrality can be quite helpful. The goal is to be prizing of all the client's experience, the painful and the happy, the destructive and the generative, all without the addition of assessing or conveying whether it is good or bad. It is a therapeutic nonattachment to what the client is saying or how they are saying it; it is neutrality. However, it is neutrality with appreciation and gratitude that is benign and accepting, not a neutrality that is clinical, cold, or unengaged. It is the steady hand of accepting whatever comes your way from the client.

## Empathic Understanding

Empathic understanding is the most teachable, practicable, and improvable of the three core conditions. It is likely that you will spend many hours in the coming weeks and years working on the skills of empathy, and hopefully integrating the attitude of empathy into your daily life and especially your time with clients. Empathic understanding is expressed quite simply by

Table 1.2   UPR to Self and Others With Synonyms

|  | Toward Self | Toward Others | Synonyms |
|---|---|---|---|
| Unconditional Positive Regard | Congruence, Self-Acceptance, Inner Peace | Warmth, Respect, Not Praising or Criticizing, Being With | Nonpossessive Love, Benign Neutrality, Being Nonjudgmental |

the commonly known phrase "walking in someone else's shoes." It is sensing another's feelings, learning another person's perspective, understanding his or her thoughts, and especially seeing how another person reacts to and shapes the experiences he or she is living through.

Also this condition has a more explicit behavioral directive than the other two core conditions. It asks you to try to let the other person know that you are "feeling them," that quality that makes them want to say "Exactly! You get me." This process of holding an empathic attitude and trying to communicate the experiences of empathy that you have inevitably leads to an empathic understanding process, a process by which your experience of empathy is improved with each attempt to show how you understand because the client corrects you, or as often corrects their perceptions, clarifying their own experience at the same time that they clarify the limits of your empathy. This process of empathizing, sharing your empathy, receiving the correction, and then learning a further client experience is a primary way that therapeutic change occurs. It is a process of clarification, or accurately symbolizing and unfolding new experiences. Neither the client nor the therapist is right. Instead, empathic understanding is the process of both the client and the therapist humbly learning and changing through this mutual reaching for the client's experience. Table 1.3 provides key points about empathy.

## Examples

You will undoubtedly practice some form of active listening, paraphrasing, reflections, or empathic responding at a micro-single response level within your training. There is no substitute for this practice and the immediate feedback you get from clients (or peers with whom you practice) and the delayed feedback you get from observers, instructors, or supervisors. Nevertheless, here are some examples from expert person-centered counselors working with people of different ethnicities from themselves. In the example, "C" stands for Client and "T" for Therapist.

Table 1.3 Empathy as an Attitude and Nonverbal and Verbal Behavior With Synonyms

|  | Attitude | Nonverbal | Verbal | Synonyms |
|---|---|---|---|---|
| Empathy | Curiously listening for others' experience, especially their reactions, feelings, and meanings | Conveyed through body language and tone of voice | Conveys precision with a self-correcting tentativeness | Communicated Understanding, Resonance |

## Barbara Brodley and Alejandra

Here is an example of empathy and unconditional positive regard from Barbara Brodley working with Alejandra (Moon, Witty, Grant, & Rice, 2011):

C3: Anyway, there was a day when I just couldn't take it anymore and I went and talked to the Dean. I was in tears. And he was receptive and he was really caring about it and so on. (T: Mhm, hmm) But anyway, what ended up happening was something very strange. Years later, a couple years later, I was in a workshop. And suddenly I see this woman. And I wanted to hide. That's the very strange thing that I'm telling you about (T: Mhm, hmm) that the reaction I had was just very odd.

T3: Very odd for you . . .

C4: Yes! Because

T4: You wanted to get away.

C5: Oh, I wanted to hide. (T: Mhm, hmm) It's not just that I wanted to avoid her. I wanted to hide.

T5: I see.

C6: It was just so weird. (T: Mhm, hmm)

T6: The feeling of hiding was something very much, in yourself. Like, hiding within you, even within yourself. Or a shrinking . . .

C7: Yeah. Shrinking is a good word for it. (T: A shrinking quality.) Shrinking is a good word. (Sigh)

T7: As opposed to simply avoiding someone, (C: Right) wanting to leave the room, (C: Exactly) or get out of the way (C: Right) much more physical, I . . .

C8: More physical, and more, um . . . Oh let me think, um well I like "shrinking." (T: Shrinking, mhm, hmm) Yes and it was as if I were this lesser person or that I didn't want to be seen. (T: Mhm, hmm) I just, like that.

T8: Almost ashamed?

C9: Yes! I guess (T: Mhm, hmm) Yes, mhm, hmm, mhm, hmm.

T9: Well, this is a very remarkable reaction, (C: Yeah) where you (C: Mhm, hmm) seeing this woman who exuded prejudice before (C: Yeah, right) brought that response.

C10: Yeah, right, exactly. I think was shame (T: Mhm, hmm). Yes. (Pause). So I thought, "Oh my God, what is this?" (pp. 360–361)

In this instance, notice how the counselor and the client overlap with each other, confirming and refining each other's understanding as they mutually explore the client's experience from within the client's perspective. Notice the use of "Mhm, hmm" and how it conveys an affirmation of understanding reciprocally. To me, in addition to acceptance of the difficult experience that they are evoking, describing, and working through together, the excerpt demonstrates that the conditions are not discrete. It is not consistent with person-centered therapy to be only empathic but not unconditional (think, manipulative used car salesman) or congruent but not empathic (think, I'm going to level with you and tell you how it really is). The key is an integration of the core conditions so that a therapist is *present* within the relationship. When all three of the core conditions are experienced to a high degree, presence emerges. In this sense, the conditions are just sides of a triangle that offer different angles on a person-centered meta-condition of presence. Figure 1.1 A and B and Figure 1.2 depict these examples of core conditions separated or integrated together demonstrating presence in visual formats.

Later in life Rogers discussed the concept of presence that has often been cited to explain this quality of a mature integration of the core conditions within the personhood of the therapist and/or within the context of a particular therapeutic relationship:

> I find that when I am the closest to my inner, intuitive self—when perhaps I am somehow in touch with the unknown in me—when perhaps I am in a slightly altered state of consciousness in the relationship, then whatever I do seems to be full of healing. Then simply my presence is releasing and helpful. At those moments, it seems that my inner spirit has reached out and touched the inner spirit of the other. Our relationship transcends itself, and has become part of something larger. Profound growth and healing are present. (Baldwin, 2000, p. 36)

A. Empathy not overlapping with UPR (e.g., Used car salesman may be empathic, but it is not integrated with unconditional positive regard toward you.)

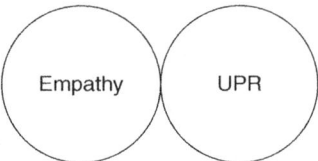

B. Congruence not overlapping with empathy (e.g., "I'm going to level with you and tell you how it 'really' is.")

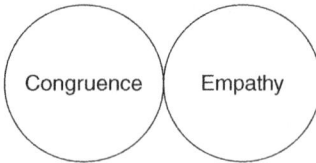

**Figure 1.1** Depictions of Separate Core Conditions

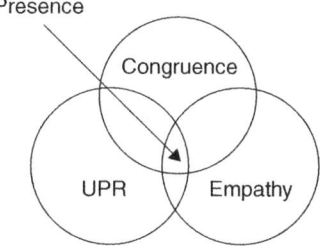

**Figure 1.2** The Integrated Core Conditions

Therefore, the practice of person-centered therapy is the practice of the core conditions such that the whole becomes greater than the sum of the three parts. The practice is one of forming, maintaining, and deepening a relationship that will allow the client to succeed, a relationship that fosters the client to be the true hero of the therapy process. The practice is more than a practice. It is a way of being in relationship.

## Summary

- The person of the therapist appears to be a key common factor.
- Rogers's (1957, 1959) Necessary and Sufficient Conditions provide 6 conditions, including what are commonly referred to as *the three core conditions*, which if present facilitate client growth and improvement.
- Congruence has several levels, such as genuineness or sincerity in the relationship and a self-acceptance and consistency between one's experience and self-concept.
- Unconditional positive regard (UPR) is acceptance, neutrality toward clients' thoughts, feelings, perceptions, with warmth and respect, a nonpossessive love.
- Empathy is understanding of a client's experience, his or her thoughts, feelings, meanings, and especially reactions in the moment to his or her narrative and experience of the therapist.
- Presence is the intersection of congruence, UPR, and empathy, a quality that is larger than the sum of the parts, a way of "being there" that is therapeutic for someone.

# 2
# Client

Life involves suffering. It is inherently difficult, and people are imperfect. Clients are just people who are struggling, typically overwhelmed and/or overwhelming those around them. In other words, clients are people like us. In fact, we counselors are often clients ourselves. About 85% of American therapists have themselves been clients, and rates are even higher in international or psychodynamic samples, where therapy for therapists is more normalized or a requirement (Norcross, 2005).

Clients have a tough job in therapy. They are trying to confront painful, difficult memories, feelings, and information to solve problems. However, clients are normally skilled to avoid discussing, thinking, and especially feeling the meanings of the very problems that brought them to therapy. They have developed behavior patterns that support this avoidance and may keep them stuck in a spiral of difficulty. Moreover, they sometimes do not connect with their therapist or do not want to be in therapy, adding another level of challenge for them.

Despite these truisms, research strongly confirms that clients are ultimately heroes and the engines for change (Bohart & Tallman, 2010; Duncan & Miller, 2000). Lambert's (1992) often quoted review found that 40% of the variance in therapeutic outcome is attributable to clients and extratherapeutic factors, 30% attributable to therapeutic relationship, 15% to placebo/expectancy and hope, and 15% attributable to specific therapeutic techniques (see Figure 1 in the Preface). More recently, Lambert has suggested that 86% of the variance in outcome is actually due to clients and their life circumstances, including known factors like problem severity, sources of help in clients' lives, economic, coping, and intelligence resources, psychological mindedness, motivation, and other unknown factors (Duncan, Miller, & Sparks, 2007). Similarly, Wampold's (2001) meta-analysis found that 87% of the variance in therapy outcomes is due to client and context, including unexplained variables. From these more recent models, only a small percentage of 1% to 2% of why change happens in therapy is actually due to treatment effects of a specific model. Therapist or therapeutic relationship factors appear about 10 times larger (10%–20%) than general or specific techniques or placebo (Duncan et al., 2007). Look again at the range of those scores:

- Client: 40%–87%
- Therapist relationship: 10%–30%
- Therapist techniques: 1%–15%

It's really not all about you. It's not even about what kind of relationship you can foster. It's really about the client.

The person-centered therapist aims to sincerely and fully engage with the client's whole person, his or her vulnerable yet heroic nature. For therapists-in-training, it is hard to just *be with* someone when they suffer, to not try to "fix." Another problem is knowing "where I end and you begin," that is, boundaries and the recognition that my experience, values, and solutions are not yours no matter how similar they may seem to be. It is the person-centered therapist's goal to *learn from* more than *teach* the one who is overwhelmed. It takes courage and patience. In a sense, a person-centered therapist's main job is to model how to face the music. To courageously confront the latent, unfolding parts of the client's experience that they typically avoid or do not accept and thereby paradoxically activate the resources rather than the deterrents in the client's life. Ultimately, person-centered therapists strive to meet person-to-person in this process. It is in this humble meeting where all parties learn that the resources of the client can often see them through the worst suffering. The tools and behaviors that were utilized to survive are no longer needed and new tools and behaviors can emerge and be refined.

This chapter explores the "client" conditions of contact, incongruence, and perception as well as the foundation on which person-centered therapy rests: the actualizing tendency. The *actualizing tendency* is the drive for people to maintain and enhance themselves. The *actualizing tendency* is inferred from the motivation emerging from client suffering with contact and incongruence, and most especially their perceptions along the way. It is about how a way of being *nondirective* represents one of the most challenging yet rewarding stances for therapists to fully encounter the client. This chapter is about fostering or valuing the client's own contact, motivation, and perception.

# Contact, Incongruence, and Perception

In the last chapter, you learned about the six necessary and sufficient conditions, especially the three core relational conditions of congruence, unconditional positive regard, and empathy. The other three are discussed here.

## Contact

Rogers discussed contact perhaps the least of any condition. It was presumed to be a precondition in a way: a client needed to be in contact with a therapist to benefit from therapy. Prouty (1994) developed, tested, and trained people in pretherapy, a variant of the person-centered approach that focused on situations where this first condition was not consistently met, such as with persons with psychotic symptoms or severe intellectual or developmental disabilities. In these cases, people may have limited contact with self, others, or world. In other words, people out of contact may not be in touch with reality, emotions, body, language, or memory and as such may not have much contact with the therapist.

Even with the most hard-to-reach people, there is more potential for contact than you expect. One can empathize at concrete levels consistent with the worldview of the client. Responding at the level of simple reality, facial, body, word-for-word, or reiterative statements can engage difficult to reach persons (Sommerbeck, 2006). The main point is to respond to and activate latent resources that can facilitate contact (see Table 2.1).

## Examples

Consider the following situation in which you are treating a client who has been diagnosed with schizophrenia. The client appears disconnected from the reality you are experiencing. The client is catatonic, not moving, and staring off into space. Maybe they say something incomprehensible or

Table 2.1 Contact Functions and Examples of Concrete Responses

| Contact Functions | Examples of Responses |
|---|---|
| Reality Contact | Situational Reflection |
| Affective/Emotional Contact | Face or Body Reflections |
| Others/Communicative Contact | Word-for-Word or Reiterative Reflections |

are silent. You feel stuck because you can't perceive contact and cannot think of what to accept or empathize with per se. Don't think. Just notice. Then reach out with what is concretely there, trusting that something will emerge that is meaningful. You might say,

- "You are sitting in a chair. You are looking out the window. The sun is shining." (situational reflection/reality contact)
- "Your arms are stiff and your head is tilted." (therapist moves arms and head to match client's posture) (body reflection/affective contact)
- "Your face looks sad. There are tears in your eyes." (facial reflection/affective contact
- Client: "OWWSH, I'm sick of zoo's fingers" Therapist: "You are sick of the zoo's fingers." (word-for-word reflection/communicative contact)
- Pause. Therapist says, "You said 'OWWSH, you are sick of the zoo's fingers.'" Client says, "The soup hurt my fingers." (reiterative reflection/communicative contact)

Once contact is reached, there is often new material that can help problem solving by the therapist and client, such as treatment for the person's burned fingers, and there is a chance to build a helping relationship that can foster more expression and possibilities.

Many students began work in counseling related to people who are out of contact, especially as case managers. I worked in residential, inpatient, nursing home, and full day treatment programs at the start of my career. In these settings, though growth may be subtle, it's easy to see how facilitating contact can help release potentials for change in even the most disconnected clients. Nevertheless, many counselors do not work with people like this. Even if not, learning basic contact techniques can help by responding to what is and/or potentially what is there rather than focusing on what is not or what is a problem can help. Contact responses are a good way to focus on client resources rather than become disheartened and lose the core relational conditions.

## Incongruence

Clients are self-motivated. Sometimes, this is not obvious. Perhaps easier to see is that clients are vulnerable and suffering and want to change. However, sometimes, this is not obvious either. Person-centered theory proposes that people have internal conflict, sometimes more visible externally (think substance abuse, personality disorder). Nevertheless, they have a sense that they are not figuring something out (in person-centered theoretical terms *symbolizing* their experience). There is a conflict between the way they think of themselves (their *self-concept*) and their *experience* (their feelings, meanings, or reactions).

Everyone experiences incongruences every day, often in subtle ways. You are implicitly helping clients sort through each of them nearly constantly. Sometimes, good empathic responses can help a client feel particularly understood when they make these incongruences explicit by making statements like

- Part of you feels sad about the loss, but another part feels relieved that it is finally over
- You really want it to happen, but at the same time you are kind of scared it might
- You think about the situation like this, but in another way, that just doesn't feel right
- You are laughing about it because it's ridiculous, but in another way it really hurts

In each of these situations, you would want to be more specific rather than generic (e.g., losing your mother, getting divorced, graduating, etc.) with your word choices, but I give them as examples of how sometimes empathy toward incongruent states might appear. In all instances, attention to the client's perception is key to whether a statement is experienced as empathic rather than patronizing, observational, confrontational, academic, or just plain "off."

Likewise, they have a drive to recover. Bohart and Tallman (1999) have gathered together extensive evidence of how clients "self-right."

## Examples

- 70% of clients report changes that occur within a week outside of therapy (Miller, Duncan, & Hubble, 1997).
- 60% of clients improve between the time they make an appointment and when they first show up (Miller et al., 1997).
- Many clients actively think through their problems on the way to the therapist's office.
- Many clients report dreams and significant events the night or morning before an appointment.
- Lambert (1997) estimates that 40% of people appear to spontaneously recover from their problems.
- People in therapy appear to use more nonprofessional support systems, suggesting that part of therapy's effects are to activate social resources already available.
- Clients sometimes re-create suggestions, or ascribe suggestions to counselors that the counselors never remember making, often with adaptations more suitable to their actual lives.

Incongruence, or vulnerability, drives people to learn and adapt, especially when presented with a facilitative relationship (see Table 2.2).

**Table 2.2** Statement on Incongruence and the Process Toward Congruence

Conflict between experiences and/or self-concept can be manifested in anxiety or vulnerability that when unconditionally, empathically confronted leads to integration or congruence.

## Perception

The core relational conditions—empathy, unconditional positive regard, and congruence—have strong research support. However, they have irrefutable research support when they are measured from the client's—not a researcher's, supervisor's, or therapist's—perspective. (It is a humbling reality that the therapist's perspective on the relationship is the worst of these to predict outcome.) In other words, whether or not the client feels or believes that you accept and understand them is what matters most. It is the client's perception of your genuine relationship with them that is golden (Cornelius-White, 2002, see Table 2.3). Each response of the therapist is an opportunity for the client to perceive your empathy and unconditional positive regard, and each response from the client is the therapist's opportunity to tailor their next response to fit the client perceptions. Therapeutic responses need a confident tentativeness. Speak not only with efficacy to communicate the core conditions, but also with tentativeness demonstrating that the client is the expert on their experience, including their experience of your therapeutic relationship.

Far too common, beginning and experienced therapists alike trust too much in their theory, experience, or intuition more than the client's perceptions—to their clients' and their own detriment. Even when presented with direct client feedback (orally or on working alliance and session rating assessments), some therapists will claim the client "was not ready to change," "had too little self-awareness," or "was not open to feedback." Therapists are sometimes just too attached to their expertise to understand or remember that it is the client's perception of the relationship that is key to therapeutic change, not their own. The best therapists actively seek feedback and change in response to client perceptions. (The same, unfortunately, is frequently a problem with counselor educators. They too forget that learning is a process fueled by the student, not the sage on the stage.) Here's a supportive nugget from research: 90% of therapists rated themselves as being in the top 25% of providers (Dew, 2003, as cited in Cooper, 2008). The following section discusses the importance of the feedback loop of assessing client perception directly to improve therapist performance.

Table 2.3  Characteristics of Client Perception

| Sixth Condition | Therapist Stance | Therapist Flexibility | Feedback Loop |
|---|---|---|---|
| Core conditions are most potent when perceived, which is fostered by | Therapist confidence tentativeness and | Smooth adaptation to what the client does or does not perceive and how they change with each perception so that | Person-centered therapy works through an interactive loop |

## Example

Jasmyne raises her voice while gesturing with her arms and says, "He disgusts me. I want to strangle him. He won't talk with me at all after he does something he knows is wrong. If I could just get him to talk, then we could work it out." You as her therapist say, "He really pisses you off with the silent treatment." You might have focused on some other part of what she said, but your empathic attitude appears to resonate with the frustration. She starts to cry and in a few seconds says, "I'm not pissed so much as disappointed. This relationship can't work if it's one-sided." You say, "Uh huh, you just wish he was in it *with* you." She says, "*Exactly*, I truly do."

In this example, it is difficult to tell whether Jasmyne found your first statement empathic or not. Chances are she did, and feeling understood on the anger, the hurt arose. However, your statement might have been off, and she didn't feel pissed so much as hopeless and at her wit's end. Either way, the feedback loop of letting her perceptions inform your empathy allowed you to adjust your next statement, whereby she clearly felt understood.

## Client-Directed, Outcome Informed, Client Hero

Listening for, and at times soliciting, client feedback is central to person-centered therapy. Duncan et al. (2009) gathered evidence and implications about "the heart and soul of change." They have also constructed a website with many useful resources for integrating their findings into practice: https://heartandsoulofchange.com. They summarize their approach as client-directed (that is, the therapist is responsive to, adapts, and/or designs treatment to fit with the client's model of change) and outcome-informed (they adjust treatments when clients do not recover within the reliable change windows they have established). Central to these ideas is utilizing simple assessments like the Session Rating Scale and Outcome Rating

Scale, four-item process and outcome measures that take under 1 minute to administer but provide immediate feedback to the counselor and (if inadequate relationship or therapeutic growth is indicated) the client. From a person-centered perspective, this is a means of formalizing feedback to help a counselor better understand and value the client's perspective. It gives more potential assurance to the *perception* condition and allows for alterations in approach as needed. The use of formal feedback instruments is also discussed in Chapter 5.

There is much attention paid to what works in therapy. There is less paid to what helps clients stay in therapy. Lambert and Ogles (2004) have found that dropout rates average about 47%. Proven techniques are not effective if the client is not actually in therapy. An element that clearly prevents client deterioration and attrition is incorporating client feedback assessments. Regardless of orientation, building in client feedback on relational elements has shown effects to reduce the amount of canceled and no show rates in the range of 25% to 40% (Claud et al., 2004, cited in Bohanske & Franczak, 2010). If clients have a strong working alliance with the counselor consistently for the first 3 weeks, then retention and therapeutic change are likely, with a higher proportion of that change occurring in the first month than later. Almost three-quarters of the time when clients are not getting better within the first month, it is because of a problem within the client-therapist relationship (Duncan et al., 2010).

Common factors discussions often include the *dodo bird verdict*, the finding that all bona fide approaches to psychotherapy are effective for outcome. However, the verdict is not necessarily true for attrition. For example, in a comparison of treatments between cognitive behavioral therapy (CBT) and person-centered therapy (PCT) with 550 clients where the treatments were found to be appreciably the same on almost every outcome measure, attrition was shown to be 40% in CBT but only 9% in PCT (McDonagh et al., 2005, cited in Wampold, 2001, p. 868).

## Actualizing Tendency

The six conditions of person-centered therapy are built on the actualizing tendency, which is manifested as a trust in the person and a trust that the client will try to maintain and enhance. When a person is in a facilitative environmental context, like a therapeutic relationship, they tend to develop and grow more obviously. However, even in the direst of contexts, the person-centered approach views persons as having potential, choice, and power to develop, even if it may look weak or perverted from an external view. Rogers (1980) gives a classic example of a potato in a basement, sprouting toward the light, potentially looking pathetic in comparison to how it might thrive under better environmental circumstances, but valiantly striving to become against all odds. In earnest, whether obvious are not, people are constantly adapting to life, solving problems and self-righting from negative emotional experiences.

Another word for the actualizing tendency is *grit*. Rogers, Lyon, and Tausch (2013) identified grit as one of the most important aspects of becoming an effective helper. Grit is determination, unyielding will in the face of difficult circumstances. Identifying, valuing, and fostering grit is a goal for the person-centered therapist, both within themselves and with their clients.

## Nondirective Attitude

A natural implication of the actualizing tendency is a nondirective attitude within the therapist. Rogers became fond of Zen and Taoism in the final decade of his life, often quoting Lao-tzu and the *Tao Te Ching*. Rogers (1980) wrote,

> But perhaps my favorite saying, which sums up many of my deeper beliefs, is another from Lao-tse:
> If I keep from meddling with people, they take care of themselves,
> If I keep from commanding people, they behave themselves,
> If I keep from preaching at people, they improve themselves,
> If I keep from imposing on people, they become themselves. (p. 42)

Brodley (2005) defines the role of the nondirective attitude in person-centered therapy thus (italics in original): "(1) It expresses *values* in the therapy relationship; (2) It is *intended to protect* the client" (p. 2). The values are trust and respect for the client and the protection is for the client's safety, self-determination, and being an accepted, effective person and protection from the therapist's structural power as an authority within the therapeutic context. She continues,

> The non-directive attitude is psychologically profound; it is not a technique . . . with time . . . it becomes an aspect of the therapist's character. It represents a feeling of profound respect for the constructive potential in persons and great sensitivity to their vulnerability. Therapy is an art. As an art, it involves freedom with great discipline. (p. 3)

In other words, the nondirective attitude is not a prescription or proscription to "Do" or "Do Not" do anything. It does not mean as a person-centered therapist that you cannot suggest techniques when a request is made or implied. It does not mean that you are passive or uninvolved. It is a deep appreciation and way of being that powerfully facilitates others without imposing anything on them.

## Conclusion

The client is the main engine of change within psychotherapy. The therapist's first goal is to understand, value, and foster that reality. The client's contact with reality, themselves, or their therapist is a precondition, while client incongruence and ambiguity, vulnerability or suffering provides sincere motivation for change. Client perception of empathy, warmth, and genuineness (the core relational dimensions of person-centered therapy) is of central importance. They are most demonstrably effective if successfully communicated and perceived by the client. Instruments advocated by Duncan et al. (2010) in their client-directed, outcome-informed approach formalize the feedback loop to convey client perception and suggestion adaptations counselors can make to improve care. I leave you with information about therapists themselves as clients because this can motivate you to take the plunge or value your past experiences in therapy. Therapists as clients not only improve their empathy and relational conditions as therapists themselves, but also help you improve the quality of your life (Norcross, 2005).

### Example: Therapist, Heal Thyself

It is clear that therapists themselves prefer to be treated and inevitably learn through therapy in the humane fashion advocated for by common factors and the person-centered approach. Bike, Norcross, and Schatz (2009) reviewed, replicated, and expanded earlier studies on therapists' own experiences of attending therapy. Sampling from thousands of therapists, they found that the six lasting lessons learned by therapists from their experiences as a client were largely person-centered themes. These themes were

- the centrality of warmth, empathy, and the personal relationship;
- knowing what it feels like to be a patient;
- the importance of transference and countertransference;
- the need for personal treatment among therapists;
- the inevitable human-ness of the therapist; and
- the need for more patience and tolerance in psychotherapy. (Bike et al., 2009, p. 26)

Even behaviorally oriented therapists rarely chose themselves to go to behavioral therapy (<10% in four different studies by different authors) or comment on the behavioral model elements about their own treatment, preferring and personally valuing the relational style of psychodynamic or humanistic counseling (Norcross, 2005). It is not that therapists of all orientations do not value the behaviors they have changed or transference connections they have made or other theory specific elements (indeed they do); it is that the elements of therapy that stick with them are relational. In addition to learning these lasting lessons through their own therapy, therapists tend to prefer humanistic dimensions in seeking therapy, particularly once experienced or when returning to therapy on a second or subsequent time in treatment (Bike et al., 2009). As you progress through the next chapters related to multiculturalism and developments and expansions from the person-centered model, I hope the loop between the relationship-client-relationship conditions sticks with you.

## Summary

- The client is the hero of the therapeutic enterprise. Research is clear that client factors are strongest in how and how much a client improves. Humility as a therapist is a virtue.
- Contact is the first of the six conditions, which may consist of connection to the world, one's self, or others and can be fostered by concrete reflections of reality, affective, or communicative content.
- Incongruence is a client's inner conflict, not only presenting often as vulnerability or anxiety, but also as a rich resource for the therapist to empathize and thereby propel client motivation and foster integration and development.
- Perception is the crucial element that shows whether therapist congruence, UPR, and empathy were successfully communicated and serves to provide a feedback loop for the therapist to adjust their responses in a way that clients can perceive the core conditions.
- Client outcome can be fostered through more deliberate requests for feedback from clients. This feedback can be accomplished with short assessments measuring their session satisfaction and improvements or deterioration so that therapy is client-directed and outcome informed.
- Actualizing tendency is the motivational force in each of us that leads one toward maintenance and enhancement; it is the foundation on which person-centered therapy rests.
- The nondirective attitude is a natural extension of the actualizing tendency, whereby the therapist does not impose on the client but aims to offer a therapeutic climate in which the client is best fostered as the hero of the therapeutic enterprise.

# 3

# Evolution of the Person-Centered Approach

Person-centered therapy is referred to by different names, such as non-directive or client-centered. One name that unfortunately has stuck around is Rogerian. Rogers himself did not like the term "Rogerian," and as Mearns (1980) explains, it "runs totally counter to the emphasis which the person-centered approach places on helping the individual trainee to develop along lines which fit his own self" (Para 1). Through its emphasis on idiosyncratic practice (Keys, 2003), person-centered therapy has been uniquely forward looking and focused on the development of itself as a theory. Rogers (1961) wrote,

> Experience is, for me, the highest authority. . . . It is the basis of authority because it can always be checked in new primary ways. In this way its frequent error or fallibility is always open to correction. (pp. 23–24)

In a similar vein, the World Association for Person-Centered and Experiential Psychotherapy and Counseling (2000) touts an "an openness to the development and elaboration of person-centered and experiential theory in light of current and future practice and research" (Para 1). Indeed, Rogers himself lead many of the first studies of psychotherapy, publishing the first transcripts and conducting innovative quantitative and qualitative designs (Kirschenbaum, 2008).

In actuality, person-centered therapy today refers to several different developments or sub-approaches, each with its own theoretical, practice, and research traditions (Sanders, 2004). In addition to many variations, the person-centered approach has been uniquely interdisciplinary beyond the typical bounds of psychotherapy, with more empirical study and applications than any other approach to therapy, including in education, management, mediation, religious studies, neuroscience, information technology, and other domains (Cornelius-White & Motschnig-Pitrik, 2010; Cornelius-White, Motschnig-Pitrik, & Lux, 2013a, 2013b).

The goal of this chapter is to introduce the reader to a variety of developments in the person-centered approach and suggest some related practical tips. Each section offers the fundamentals of the development, or sub-approach, followed by some of its contributions to a common factors approach to therapy. This chapter includes discussion of the following research and practice approaches: classical client-centered, focusing, intersubjective, and emotion-focused. It concludes with two additional developments: interdisciplinary research and practice and the futurism of the person-centered approach.

## Classical Client-Centered or Nondirective Therapy

### Fundamentals

While most counselors might argue that empathy, unconditional positive regard (UPR), and genuineness are necessary but not sufficient, classical client-centered therapy emphasizes the "necessary and sufficient" conditions as just that. Advocates point out the empirical basis for *client perceptions* of these three variables are demonstrably effective (Cornelius-White, 2002; Patterson, 1984; Stubbs & Bozarth, 1994). Bozarth (1993) goes so far as to say the research shows that they are *not necessary but they are sufficient* by adding that clients often make beneficial changes in the presence of therapists with low levels of the conditions or in self-help settings without a clear relational context.

Other advocates argue that research is irrelevant or of secondary importance. They assert that the basis for helpful interactions between people is ethics or morality (Grant, 1985; Schmid, 2013). This view, presented with secular or religious foundations depending on the author (especially Christian, Buddhist, and/or Taoist), emphasizes that it is not the strategic value of nondirectivity that matters, but that interacting with sincere, non-possessive attention is the crux. To many classical client-centered practitioners, nondirectiveness is not an instrumental practice, that is, you are not nondirective for its effects, but a principled practice, that is, because it's the best way to treat people (Grant, 1990). One of the first and most enduring statements on nondirectivity is a statement of Nat Raskin, which Rogers (1951) summarized:

> He (or she) tries to get within and to live the attitudes (of the client) expressed instead of observing them, to catch every nuance of their changing nature; in a word, to absorb himself (or herself) completely in the attitudes of the other. And in struggling to do this, there is simply no room for any other type of counselor activity or attitude. (p. 29)

The classical view reveals the immediacy and vitality of a person-to-person encounter. When you are truly absorbed, you are quite simply only in flow rather than your head, theory, or distractions. You and the client experience a therapeutic way of being.

In summary, the classical view emphasizes a full trust in the actualizing tendency (the client as change agent) and suggests that a nondirective attitude pervades the practice of person-centered therapy, whereby only empathy, congruence, and UPR have any place. However, in what often seems a paradox to people less familiar with the actual practice, classical client-centered therapists are usually quite free in allowing for spontaneity and variation in their behavior, especially if lead by the client. These "experiments" involve behaviors that deviate from the core conditions in the moment. They might spontaneously offer a perspective or activity that might look like something from another theoretical perspective. Likewise, at times a classical client-centered therapist may speak from his congruence about his or her own experience if a reaction is persistent and perceived as also somehow empathic or unconditional. Classical client-centered therapists do (infrequently) self-disclose. In this sense, the classical client-centered therapist can still be quite flexible, but only as is perceived to be in tune with the client's experiences. Classical client-centered therapy developed from the late 1930s to the 1950s but is still practiced with passion and success today.

## Common Factors Connections

In classical client-centered therapy, the client is THE therapeutic change agent. The relationship is the context in which the client can most likely succeed. These are quite simply the focus of classical client-centered therapy. The relationship is defined strictly by a nondirective attitude, which is seen as an implication of purely, and only, practicing according to the core conditions as perceived by the client. The therapist is rarely concerned with the specific outcomes of therapy, instead being focused on the practice of the core conditions. The attitude and way of being is more important than the technique. Indeed technique is devalued because it smacks of a therapist being "up to" something, or presenting themselves as the authority rather than fully and sincerely being absorbed in the client's experience. Because the attitude is the quintessential element, there is room for spontaneity as guided by the client or by personal expressions of the therapist that are rooted in unconditional empathy and self-awareness.

## Example

Classical client-centered therapy offered the very first transcribed sessions of any approach to therapy (Kirschenbaum & Henderson, 1989) and more recent advocates have been at the forefront of cataloguing and making publically available such sessions, particularly the work of Carl Rogers himself and Barbara Brodley (e.g., PCAYorks, 2014). Here is a short interaction that starts a session of Carl Rogers and Dione from a filmed session called "The Right to Be Desperate" (Brodley & Lietaer, 2006), which demonstrates classical client-centered therapy.

### *Carl Rogers and Dione*

C 4: I think that I've listened for so long to other people about who I was (T: Mhm, hmm) and I remember in 2nd grade I was a potential credit to my race. That was one of the . . . (T: Mhm, hmm) I used to wonder why I couldn't be a credit to somebody (T: laughs) else's race also. I think I was really conditioned to be something. To be some kind of a symbol, or (T: Mhm, hmm) whatever, and not really being a person, you know, I kind of missed out on my childhood to an extent. (T: Mhm, hmm) I don't really regret it. I don't think I regret it anyway, but I've really been through a lot of changes. And I think that now, after finding out I had leukemia and after dealing with the leukemia in the way I did, it's just really incredible, you see. It was last June when I found out, and um, I proceeded to get everything in order because I was told that I had less than a year to live. (T: Mhm, hmm) And uh, that was the trip. (T: I'll bet.) That was a trip.

T 4: A trip into a fairly dark place, I suppose.

C 5: Oh, yeah. For sure. For sure. On one hand I accepted the death. I, you know, at my young age I think I've lived long and a great deal. But that was the start of some things that really has had an effect on me today. (T: Mhm, hmm) I'm much happier than I've ever been . . . today. (T: Hmm) I'm much happier. (T: Hmm) But there's a lot of hurt too. There's an awful lot of hurt and I think I'm just beginning to realize that. Um, because you know, in being a credit to your race, in being an outstanding student, an outstanding scholar, an outstanding football player, whatever; it leaves you little room to be . . .

T 5: You've been meeting other people's expectations of you and it seems that that's what you should do now. I guess you're really questioning that very much.

C 6: Oh, yeah. Tremendously.

What do you notice about how Carl responds? How does the response express his inner experience of empathy? Can you tell that the client's "point" and the client's own reactions to his narrative are the targets of Carl's response? Notice the timing and economy of words.

# Focusing

## Fundamentals

Rogers's most influential student and colleague was Eugene Gendlin. In the 1950s and 1960s, he created *focusing* and is the "father" of experiential psychotherapy. It grew out of the process of what many successful clients were analyzed in research studies to do in person-centered therapy. These experiential learners found a way to reflect and learn in a highly engaged, sensitive, meaningful way. Focusing can be done with a therapist or alone. It is the practice of pausing life to allow unfolding of new meaning and possibilities. Focusing can be described with six steps: clear a space, perceive your felt sense, get a handle, resonate, ask, and receive (Gendlin, 1982). A short form to guide and explain focusing is available online (Focusing Institute, 2014).

## Common Factors Connections

One of the largest themes of common factors is that the client has the potential resources for change. Focusing suggests that your aim is to help them listen to their inner voice to find their intuition. Appreciate silence and moments of patience and invite them to do the same. Locate feelings in the body, explore it, and watch meaning unfold in unexpected ways. Have a sense of gratitude for the small learnings that often lead to large ones. The therapeutic relationship in focusing remains in the most important supportive role, but interventions may include inviting or suggesting techniques to help the client find their own handles to explore and learn from. The connection between mindfulness, arguably another common factor, and the person-centered approach are explored in Chapter 4, but some parallels between focusing and mindfulness are probably obvious to the reader. Mindfulness-based interventions have been shown to have a variety of benefits (Brown, Marquis, & Guiffrida, 2013; Davis & Hayes, 2011).

## Example

The following example is adapted from Gendlin (1996) cited in Focusing (2014):

> Take a moment to relax. Pay attention inwardly, in your body, perhaps in your stomach or chest. Now see what comes *there* when you ask, "How is my life going? What is the main thing for me right now?" Sense within your body. Let the answers come slowly from this sensing. When some concern comes, DO NOT GO INSIDE IT. Let there be a little space between you and that. . . . Wait again, and sense.
>
> From among what came, select one personal problem to focus on. There are many parts to that one thing you are thinking about—too many to think of each one alone. But you can *feel* all of these things together. Pay attention there where you usually feel things, and in there you can get what it feels like. Let yourself feel the unclear sense of *all of that*.
>
> What is the quality of this unclear felt sense? Let a word, a phrase, or an image come up from the felt sense itself. It might be a quality-word, like *tight, sticky, scary, stuck, heavy, jumpy*, or a phrase, or an image. Stay with the quality of the felt sense till something fits it just right.
>
> Go back and forth between the felt sense and the word (phrase, or image). Check how they resonate with each other. See if there is a little bodily signal that lets you know there is a fit. To do it, you have to have the felt sense there again, as well as the word. Let the felt sense change, if it does, and also the word or picture, until they feel just right in capturing the quality of the felt sense.
>
> Now ask: What is it about this problem that makes this quality? Make sure the quality is sensed again, freshly, vividly (not just remembered from before). When it is here again, tap it, touch it, be with it, asking, "What makes the whole problem so _____?" Or, "What is in *this* sense?"
>
> Be with the felt sense till something comes along with a shift, a slight "give" or release.
>
> Receive whatever comes with a shift in a friendly way. Stay with it a while, even if it is only a slight release. Whatever comes, this is only one shift; there will be others. You will probably continue after a little while, but stay here for a few moments. (para 7–15)

# Intersubjectivity

## Fundamentals

All forms of person-centered therapy are intersubjective, which means that they are based on the person-to-person encounter. However, several theorists and trainers have emphasized dimensions of this with concepts such as *relational depth* (Mearns & Cooper, 2005), *dialectically substantial and relational* (Schmid, 2013), and a web of inner, interpersonal, group, and systemic relationships (Barrett-Lennard, 2013).

Central to each of these variations is that a person-centered therapist is not responding only to the client's individuality, but also to the space between the client and the therapist, a space that is sometimes explored through the experience of the therapist. Schmid sees the person as simultaneously an individual (substantial) and a relationship. This means that a person's boundary and identity is autonomous yet interconnected. It is the dialectic, the contradiction or tension between these realities, which helps explain how the person-centered approach works. Mearns and Cooper (2005) have highlighted the implications of this idea as a practice of sharing one's own experience of the client to avoid negating the therapist's self through the self-protections of the client. This provides therapists a unique discourse of difference within a setting of empathy and acceptance. They view expressions of congruence and immediate encounters, as seen very commonly in person-centered groups, as important for a depth and sincerity of a therapeutic relationship. They see person-centered therapy as inherently intersubjective, which is sometimes viewed as a contrast to classical nondirectivity, emphasizing a focus on the subjectivity of primarily the client. In a less confrontational view, Barrett-Lennard views intersubjectivity more broadly. He sees the person as a web of relationships. A person is not an individual so much as they are a set of communication processes within and between close and more distant persons. In this sense, a sense of community and empathy to the person's relational field is as important as understanding their inner conflicts. Persons are always more than individuals.

## Common Factors Connections

Intersubjective person-centered therapists view the person, both the client and the therapist, as occupying a space between each other as much as within their individual selves. It is like a Venn diagram, which shifts constantly as experiences are shared and permeable boundaries shift to include meaning or emotional spaces formerly occupied by one person or the other. While intersubjective person-centered therapists vary in the style and degree of confrontation or expressions from one's own experiences, they each suggest that to live out the person-centered approach in therapy means to live out a concern for the meaning of the relationship between the client and therapist, not just providing relationship conditions as a soil from which a client's seed of the actualizing tendency sprouts and grows. The practice and relationship exists within a discipline of the core conditions such that an expression of one's own experience should be influenced by and followed with a focus on unconditional acceptance and empathy for the client's reactions to the therapist. Likewise, intersubjective person-centered therapy emphasizes that individual therapy alone is a step in the unfinished journey toward a more mature elaboration of the person-centered approach. From an intersubjective person-centered perspective, group, community, and cross-cultural dialogue and even political activism are inherent to the notion of the person (Proctor, Cooper, Sanders, & Macolm, 2006; Schmid, 2013). This has implications for how a therapist would respond in any setting. It also suggests a need for political activism and articulation of multicultural differences and interconnection, as discussed in Chapter 4.

## Example

Here is an example of an intersubjective interaction. Jim, a client, is talking to Jane, his therapist, about how his girlfriend is warmer to his friends than to him, saying, "She hugs my friends when she greets them but me, she just says Hi." Jane replies, "You don't like how she is warmer to your friends than you." Jim replies, "Yeah, makes me wonder what I did to piss her off or if she still likes me anymore. Then if I say anything, she just goes cold. She won't say a thing, reply to my texts, or anything." Jane replies, "You feel rejected. Jim, I feel for you. I'm hesitant to say this, but I wonder what it feels like right now if I say that I have heard you talk about how she is cold to you many times every week and then you reach out and she gets even colder." Jim responds, "Well it surprises me. I like to think she and I have a good relationship, but I just don't know." Jane says, "It seems like you are wondering based on what I said and you are not sure it fits. Perhaps I went too far." Jim, "No, really that's exactly what I wish she would do. I wish she would have the courage to confront things."

In this example, what do you notice about Jane's empathy and the relational, interpretive, or circular nature of it? How do you evaluate her UPR? In what way is she or Jim congruent or incongruent? How does Jim or Jane together make the interaction simultaneously about both the therapeutic relationship, and Jim's relationship with his girlfriend?

## **Experiential, Emotion-Focused**

### Fundamentals

The most current, sophisticated, and growing body of efficacy research on person-centered therapy concerns a branch of person-centered experiential therapy known as *emotion-focused* (Greenberg, 2010a, 2010b). Emotion-focused therapy is an approach that grew out of the evocative function of the therapist (Rice, 1974), focusing (Gendlin, 1982), and an integration of tasks presented in an invitational way to clients. It is process directive, meaning that it aims to suggest or instruct the client to process in particular ways, but is usually nondirective about the content.

Like much person-centered therapy, emotion-focused work not only intends to evoke and prize the feelings and meanings of the client, but it also utilizes an assessment of emotion types and markers. Three forms of emotion are distinguished: primary adaptive, primary maladaptive, and secondary (Greenberg, 2010b).

The two primary types are characterized by clients' basic, direct, and immediate reactions to a situation. We are happy when something we want occurs or sad when we experience a loss. Primary adaptive emotions tend to be linked to useful information. They tend to inspire and organize action. When a therapist responds to or empathically evokes primary emotion, clarity and action often result. Primary maladaptive emotions are old, resistant to change, and hauntingly familiar. When a client experiences maladaptive primary emotion, the challenge (or benefit with a more unconditional positive regard lens) is to tolerate, regulate, and transform them. Fundamental insecurity, long-term loneliness, or worthless feelings are examples of maladaptive primary emotions, feelings that plague one for years, or often a lifetime (Greenberg, 2010b).

Secondary emotions are the emotions that one experiences in response to thoughts or to primary emotions, such as being ashamed of a desire or anger in response to a feeling of hurt. Secondary emotions are often reduced through empathy as it facilitates us to become more aware of and accepting of our primary emotions. Exploration of secondary emotions helps generate better understanding of emotions in general, but also the thought processes and experiences to which they are connected (Greenberg, 2010b).

Markers are client statements or actions that alert the counselor of an issue with the potential effectiveness of a specific task. The counselor can invite the client to a specific intervention while maintaining an integration of unconditional empathy. Nevertheless, such invitations are prescriptive and directive, creating a conflict for classical person-centered practitioners. Examples of markers include vulnerability, withdrawal difficulty, unclear feelings, narrative retelling, self-evaluative splits, or unfinished business. Markers might signal the need for an empathic exploration, a dialogue about the alliance, focusing, or an activity less common in other forms of person-centered therapy, such as the two-chair or empty chair enactments. Each task is applied in a permissive manner that invites the client and returns to a more nondirective state if the client declines. Tasks are designed to help the client better articulate, explore, and resolve the emotions and meanings accompanying each marker (Greenberg, 2010b).

## Common Factors Connections

Emotion-focused therapy is more akin to approaches outside of person-centered therapy in the sense that it is more directive and involves conceptualization and application of specific techniques to induce change. However, it is clearly linked within the approach and often long-term study of basic person-centered practice is necessary to become competent. It is supported by unusually strong research design and positive results. Also, contributions like the central role of emotion in meaning making, inaction, and action are unique compared to other theories. It is integrative because it brings together aspects of other branches of person-centered therapy along with approaches from other models. However, like other person-centered therapies, the practice and quality of empathy, unconditional positive regard, and genuineness is crucial to its success. Relationship formation, return to core conditions when invitations are declined, and application of each task are conducted from a skilled empathic and unconditional stance. Learning to recognize process markers or emotion types and how this may influence the way in which you are empathic with or without the introduction of the prescribed tasks can be a rich way to deepen your practice of relationship and admiration for clients' suffering and resilience.

## Example

Stephanie, a client, and Rebecca, a therapist, are discussing Stephanie's loss of her father and how she continues to be frustrated that she couldn't make things right with him before his death. Rebecca is responding in an empathic following with hmm-mm's, yes's, and empathic responses to which Stephanie replies things like "yeah and . . ." and "exactly." After a few minutes, Rebecca asks Stephanie,

Rebecca: Would be willing to try something that might help you find more closure?

Stephanie: Um, Ok.

Rebecca: (after pulling a chair over) Pretend that your father is right there. Could you talk to him?

Stephanie: (sounding uncertain) How would I do that?

Rebecca: It feels awkward, and you are wondering how you might try.

Stephanie: (After a few moments, she begins crying.) Dad, I miss you so much.

Stephanie continues to say things to her father in the empty chair that she had been wanting to say for years, demonstrating a lot of conflicting emotions. Rebecca replies empathically to each thing Stephanie says and helps her get out her feelings. Stephanie starts to differentiate between feelings she was harboring toward her dad, like anger and hurt, and feelings she has had about those feelings, like guilt. She also talks about how she sometimes feels she is never good enough not just for her father, but for anyone, even herself.

In this example, how does the introduction of the empty chair help or hinder Stephanie's process? How might you categorize Stephanie's feelings? How might you phrase the invitation to empty chair that Rebecca posed to Stephanie?

## Interdisciplinary Research and Practice

### Fundamentals

As long as person-centered therapy has been practiced and researched, other applications and research related to the approach have occurred. Rogers was uniquely concerned about the interconnection between life outside the session room and therapy as compared to any other major theorist. People have applied his central interpersonal ideas to a wide variety of applications. Research from an array of sciences has been used to explain and elaborate the person-centered approach. Figure 3.1 depicts the enormous scope of the person-centered approach, inclusive of but much more expansive than psychotherapy.

As a recent example, Cornelius-White, Motschnig, and Lux (2013a, 2013b) have brought together over 50 authors from all populated continents in two volumes concerned with interdisciplinary research and applications. Some of the more exciting interdisciplinary work of the person-centered approach concerns its links with other meta-theories, such as game theory (Fisher, 2013), interdisciplinary systems/chaos theory (Kriz, 2013), and neuroscience (Lux, 2013). For example, Lux offers a conceptualization of neuroscience and the person-centered approach based on the concept of a circle of contact. He shows how neurobiological level systems are activated and/or released and others are deactivated and/or stuck in a therapeutic relationship characterized by UPR and empathy both within the therapist and the client.

- Mirror systems are activated.
- The amygdala is less activated.
- Sympathetic nervous system is deactivated.

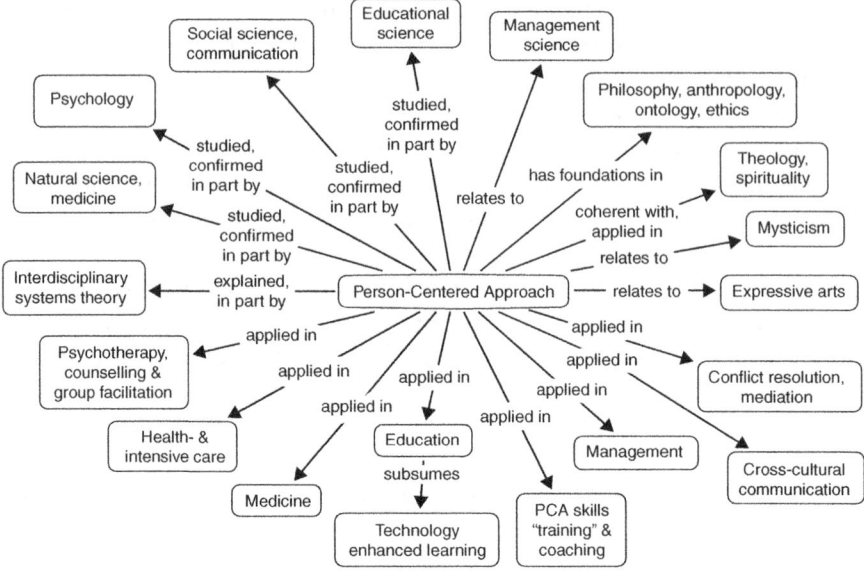

Figure 3.1 The Scope of the Person-Centered Approach
Source: Cornelius-White, J. H. D., Motschnig, R., & Lux, M. (Eds.). (2013). *Interdisciplinary applications of the person-centered approach*, p. 236. New York: Springer. With kind permission from Springer Science+Business Media B.V.

- Affect regulating structures are activated.
- Oxytocin is increased.

These changes at the neurobiological level are accompanied by an increased ability to relate to each other, reductions in anxiety, and increases in trust and expressiveness.

There are many levels to the person-centered approach, and research is accumulating at each. Figure 3.2 provides a model of how the approach is manifested and researched at these levels. The interdisciplinary applications and research provide unique bridges to explain how the approach works and suggest modifications in contextually specific situations in light of the findings of other fields.

### EXAMPLES AND QUESTIONS TO LEARN WITH

- Rational self-interest is the guiding principle of game theory. Game theory explores how you solve many large problems of today that might be termed tragedies of the commons—where individuals acting rationally deplete a common resource to everyone's detriment, such as sustainability concerns or global warming. How might the person-centered approach help (Fisher, 2013)?

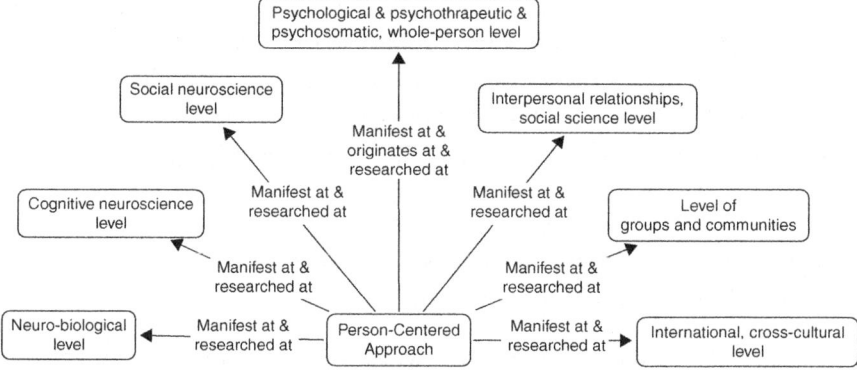

**Figure 3.2** The Multiple Levels at Which the PCA Is Studied and Present

*Source*: Cornelius-White, J. H. D., Motschnig, R., & Lux, M. (Eds.). (2013). *Interdisciplinary applications of the person-centered approach*, p. 247. New York: Springer. With kind permission from Springer Science+Business Media B.V.

- Mechanistic science has failed to explain many self-organizing phenomenon. Take, for example, when and for how long will an audience clap after a concert or speech? How does "the wave" at a football game develop and move around a stadium? How might the person-centered approach help explain this? (Kriz, 2013)
- When you taste ice cream or a favorite cookie, certain neurons fire to let you know that "yummy" sensation. What happens if you watch another person have ice cream or that cookie? Aside from secondary reactions like jealousy or asking if there is more where you could have some, there is a primary experience that is shared by the two of you. You momentarily share that experience with the other person, albeit slightly different as is unique to you. Mirror neurons fire, signaling a connection between the two of you and providing a fundamental biological basis to empathy. How does this example of neuroscience build from or elaborate on the person-centered approach? How interesting is it to see how concordance between the heart rate and skin conductivity of you and your client follow clear patterns when counseling is going well or not well? (Lux, 2013)
- What do you think happens to the heart rates, breathing, skin temperatures, and speech patterns in therapeutic relationship with a high perception by the client of empathy, UPR, and congruence?

Any approach to current psychotherapy exists in relation to all science and meaning. Too often psychotherapy theories assert an attempt to explain the truth or superiority of the approach through argumentation, advocacy, or research without contextualizing them in the larger bodies of science and meaning. Likewise, psychotherapies have often existed too much in a vacuum, not allowing a flow in and out of one approach into other sciences, practices, and meaning systems. At an individual level, every practitioner of any theory exists within a world. They surf the

Internet, watch TV, listen to the radio, read, discuss, and practice research, religion, and/or disciplines beyond therapy. The interdisciplinary history and advancements of the person-centered approach in building bridges to and from the theory provide an example for how to integrate and be influenced by the larger world in a systematic, creative way.

# The Futurism of the Person-Centered Approach and Its Implications

Carl Rogers was a futurist. While renowned for his positive view of persons and calm soft-spoken presence, heralded as the quiet revolutionary, he was actually fairly outspoken, bold, and even apocalyptic in his thinking about the future. He seemed to believe that without the integration of fundamental relationships qualities like empathy, acceptance, and genuineness into daily life from the smallest to the largest interaction, the planet was doomed (Rogers, Lyon, & Tausch, 2013). In 1922 at the ripe age of 20 years old, Rogers traveled around the world interacting with thousands of people different from him on a peace mission that was to forever change the course of his life (Rogers & Cornelius-White, 2013). During the 1940s, he inspired research on how fascism and the dehumanizing attitudes that drove World War II could be altered through core relational processes (Cornelius-White & Harbaugh, 2010). In the 1960s, he described "the person of tomorrow" who would be necessary to create a sustainable future and a new global society. In the 1980s, he conducted encounter groups in the world's hot spots of interpersonal and cultural conflict, like South Africa, Northern Ireland, and behind the iron curtain of the USSR. Throughout his life, he was looking to the future, trying to confront difficult realities that threatened all humanity, with a humane approach.

O'Hara (2003) has described how the person-centered approach offers a means to a brighter future not just for psychotherapy, but for the world at large. She wrote,

> Many now believe that humanity may have reached a tipping point—its own impasse where the challenges have outstripped our capacity to deal with them from within the worldview that created them. The impasse offers both threat of disintegration and opportunity for transformation. (O'Hara, 2013, p. 226)

### EXAMPLES AND QUESTIONS TO LEARN WITH

Following Rogers's propositions on the person of tomorrow, O'Hara (2003) suggests many themes that are suggested by the person-centered approach, which offer a guide to the future of psychotherapy and beyond:

- The importance that every voice be heard and every person recognized
- The emphasis on process—not only what is done but how it is done
- The importance of nonrational and emotional modes of consciousness and action
- The glimpse of the universal expressed in stories that are personal
- The importance of tolerating ambiguity instead of rushing to clarity and closure
- The willingness to acknowledge feeling and permitting one's vulnerability to show
- The need for communities that will respect each one as persons and where cooperation is favored over competition
- Expanding the circle of empathy to include those with whom we disagree
- The need to respect the natural and social world around us and be mindful of the balance between human activity and the natural world
- The recognition that when all is fluid there can be no preset plans, only a sense of direction and willingness to be open to feedback to change and learn as we go
- The recognition that in chaotic human situations it usually matters much more who one is than what one does (pp. 71–72)

In what ways do you think you will encounter these themes in learning the practice of person-centered therapy, in becoming a better counselor, and perhaps a person of tomorrow?

## Summary

- Classical client-centered therapy prioritizes the nondirective attitude and the core conditions.
- Focusing is an approach that can help clients process at deeper levels associated with better emotional awareness and wisdom.
- Intersubjective person-centered approaches emphasize the encounter nature of the interaction, at times offering immediacy, self-disclosure, or a systemic viewpoint within an empathic, unconditional climate.
- Emotion-focused therapy integrates an understanding of emotions and techniques from related approaches like gestalt therapy.
- Interdisciplinary approaches offer connections with a variety of fields of study and practice, expanding the opportunities for learning between approaches and applications beyond the conventional therapy room.
- Person-centered approaches have been uniquely concerned with the development of persons, including an appreciation of cultural, group, and global and neuropsychological processes that are broad and forward looking in nature.

# 4

# Multiculturalism

The goals for the person-centered therapist remain remarkably the same regardless of whom they are working with. The therapist is aiming to be sincerely empathic and accepting. However, how those goals are implemented may look quite different depending on the context. Person-centered therapy was created in and is a proponent of multiculturalism, defined here as a peaceful appreciation, coexistence, and beneficent reciprocal influence between persons of various diversities (Cornelius-White & Godfrey, 2004). These elements include race, ethnicity, gender, sexual orientation, religion, disability, and economic class. To improve multiculturalism in one's self, context, and therapy work, competencies often need to be developed, at times unlearning as much as learning.

The American Counseling Association has endorsed that multicultural competencies involve awareness of one's own culture and biases, understanding of clients' worldviews, and practicing culturally appropriate intervention strategies (Association for Multicultural Counseling and Development, 1992). The first two relate closely to core relational conditions: congruence involves awareness of prejudice, and understanding of clients' worldviews involves empathy. They both are also helped by learning and living specifics of culture through person-to-person interaction with real people who are different than you. The third is a bit more complicated because both the common factors and person-centered approaches point to the *specificity myth*, the largely unsubstantiated belief that specific disorders require and are best treated by specific interventions. However, at a basic level, appropriate interventions are those that involve empathy and unconditional positive regard, providing therapist responses and tasks when clients request or are implied by the client's worldview and confirmed by the client. As discussed in Chapters 1 and 2, the person-centered therapist is free to respond cooperatively to requests from the client for specific interventions or may spontaneously suggest them if persistently experiencing the core attitudes (Cornelius-White, 2003, 2005).

This chapter will encourage the reader to engage with multicultural issues through exploring some central concepts in multicultural competence through a person-centered lens, exploring social justice advocacy competencies, developing an appreciation of the role of multiculturalism in the history of the person-centered approach, considering ethical and religious concepts of the person-centered approach, and understand the paradoxical model that the person-centered approach represents with regard to multiculturalism.

## Central Concepts in Multicultural Counseling Competence

Racism, sexism, and other "-isms" are realities of our world. People experience a lot of cultural trauma due to their cultural statuses, which may be inherited or in some way a life choice. One can argue there has been a decrease in overt discrimination and blunt bigotry during readers' lifetimes, though others will counter that dramatic examples continue. Langton, Planty, and Sandholtz (2013) report that Bureau of Justice Statistics on violent hate crimes show a complicated picture during the last decade. For example, there was a doubling of the percentage of hates crimes that were religiously motivated from 2003 to 2006 compared with 2007 to 2011, and drops in racially motivated hate crimes, but overall the number of hate crimes did not change (Langton et al., 2013). Perhaps more insidious and widespread, subtle, covert, aversive, and institutionalized discrimination and oppression continue (Sue & Sue, 2012). Such bias, misunderstanding, and actions are often explained with concepts such as privilege, power, stereotypes, and micro-aggressions.

### Congruence: A Cultural Concept

Congruence involves awareness of feelings, thoughts, and stories about one's self, stories that are often influenced by cultural variables in subtle ways (Cornelius-White, 2006). Rogers (1961) himself discussed "subtle ways of communicating" "contradictory messages" in cross-cultural exchanges, suggesting its relation to congruence (p. 51). Schmid

(2001) wrote, "You cannot reflect on being congruent if you don't experience and consider diversity" (p. 218). Merry (2001) offers an example:

> A therapist, whose self-picture incorporates the notion that he or she is entirely free of prejudice, would experience some level of anxiety when confronted by a client of a different ethnic group . . . [and] would find difficulty in allowing prejudice feelings into awareness. (p. 179)

Hence at the most basic level, culturally congruent therapists are aware that they have a worldview stemming from their own circumstances and choices, which necessarily includes *stereotypes* (Cornelius-White, 2007b).

*Stereotypes* are commonly held, relatively fixed, and often oversimplified and under-analyzed ideas about particular cultural groups or persons representing those. Stereotypes often serve socially to reinforce power that accompanies privilege or are unpleasant and therefore avoided, not easily allowing consideration or deconstruction. Who wants to talk about ongoing stereotypes like, "black people are stupid and lazy" or "women are either mothers or whores." It doesn't necessarily matter whether these are views that people would own as their own; it matters more that they are pervasive in media, literature, institutions, and throughout social settings and continue to influence common understanding (Johnson, 2006). Stereotypes often form the basis of biases and prejudices where a person is valued or devalued due to a diversity element without consideration of the whole person or consideration of the underlying stereotype that doesn't hold water once examined. Teachingtolerance.org provides a window into understanding stereotypes and provides a wealth of free resources to improve one's cultural self-awareness as well as pedagogical kits and activities for preschool, elementary, and secondary educational settings, such as the following example.

## Examples of Stereotypes

Consider each of these examples and what mental images they evoke (adapted from Lockhart & Shaw, 2013).

- Male and female
- Sons and daughters
- Jack and Jill
- Romeo and Juliet

Now reverse the order and notice how your mental images change. What happens if you say "FEMALE and male," for example? Does a sense of priority or importance or potency or activity or responsibility change? What stereotypes are built into common language uses and how do you

Table 4.1  Cultural Congruence

| Subtle Forms of Communication | Getting Feedback | Examining Widely Socialized and Personally Held Stereotypes |
|---|---|---|
| Use of language (e.g., jargon, slang, or conventions like "he" for person), power structures (e.g., titles, time, location), or avoidance of topics | Watching for and learning about contradictory messages, especially on nonverbal communications like proximity, touch, voice tone, body language | Examining stereotypes with various groups and as triggered by particular people and working through them for deeper maturing |

hold these without necessarily realizing it? In three of these four examples minors are involved; how might all oppressions be related to society's treatment of children (Rice, 2013)? Table 4.1 describes subtle communication, feedback, and self-examination issues related to cultural dimensions of congruence.

# Understanding Client's Worldview: Cultural Elements to Empathy

*Privilege* is a concept crucial to helping many counselors-in-training and counselors to develop awareness not only of themselves but also of others who are different from them. There are advantages and immunities of people with powerful cultural status (those associated with money, positive media representation, political, corporate, and other positions of power, etc.). For example, white privilege is held if you are perceived as white, male privilege if you are perceived as male, heterosexual privilege if you are perceived as heterosexual. These privileges operate outside typical awareness if you have them but are generally obvious to those who do not. Peggy McIntosh (2013) provided a list of privileges she had as a white woman in 1988 that is representative of many pieces written on the subject.

## Examples of White Privilege

- Turn on the television and seeing people of your race widely represented, and frequently in a positive or human light.
- Go to a meeting and not have your lateness be attributed to your race.
- If you need medical attention, be assured your race will not work against you.
- Go shopping without likely being followed or harassed.

Table 4.2  Communicated Understanding Client's Worldviews Through Privilege

|  | Indirectly Helps Communication | Directly Helps Understanding |
|---|---|---|
| Engaging the Concept of Privilege and How It Applies to You | Helps you understand ways in which you may communicate that are not consistent with a client's worldview | Helps you understand a worldview without that privilege if your client doesn't have it |

- Feel welcomed and "normal" in public life and institutions.
- Ask to speak to the person in charge and likely be presented with a person of your race. (adapted from Peggy McIntosh, 2013)

Think of another example of privilege right now. What work on privilege could help you better empathize with subtleties that might be present when a client who does not hold a privilege you hold tells a simple story like watching TV, going to a meeting, or shopping?

*Privilege Lists* is a sassy, accessible website that has been inspired by McIntosh's classic list and provided examples of Masculine Dude ("Bro"), Male, Middle-to-Upper Class, Christian, Heterosexual, and Cisgender privileges (Killerman, 2013). *A List of Privilege Lists* is another web source with several groups of privileges to consider, including some unusual or provocative ones (Ampersand, 2006). Reflecting on one's privileges is a means to help understand not just one's self but also others who may not have those privileges (see Table 4.2).

# Culturally Appropriate Interventions and Unconditional Positive Regard

*Microaggression* is a multicultural concept that explores how specific interactions between people who are culturally different involve small, nonphysical actions of aggression. Sue et al. (2007) defines them as "brief and commonplace daily verbal, behavioral, or environmental indignities, whether intentional or unintentional, that communicate hostile, derogatory, or negative racial slights and insults toward people of color" (p. 271). More generally, microaggressions include subtle insults and dehumanizing implications toward anyone in relation to diversity elements of that person. Like with stereotypes and privilege, microaggressions operate only with partial awareness to many people.

Most counseling texts focus on interpersonal interactions as ways to show microaggressions, and certainly this is a worthy focus. There are well-documented cases of microaggressions in the interpersonal interactions between counselors and clients and counseling supervisors and supervisees (Constantine, 2007; Constantine & Sue, 2007; Sue et al., 2007). However, many microaggressions are environmental (Cornelius-White, 2007d). For example, Houston and Los Angeles are known as some of the most diverse cities in the United States and are sometimes revered for their thriving ethnic communities, including Latino, African American, and Asian communities in particular. However, Lipsitz (2005) reports that 100% of the dumps in Houston are located in African American neighborhoods and in Los Angeles, twice as many African American and Latino children live in areas with highly polluted air than do white children.

Appropriately intervening from a multicultural perspective within the person-centered approach involves not committing microaggressions and being aware of how microaggressions may affect the daily life of clients. Aside from answering specific or implied requests from clients with directive interventions as discussed in the earlier chapters, learning to improve your unconditional positive regard through an understanding of microaggressions can help (see Table 4.3).

## Examples of Microaggression

Consider these sentences and notice microaggressions in relation to various diversity elements inherent in them (adapted from Lockhart & Shaw, 2013).

- Mrs. Smith looks remarkably good for her age.
- I'm just a person.
- Confined to a wheelchair, Mr. Garcia continues to live a productive life.
- May I speak to Mr. or Mrs. White?
- Our founding fathers carved this great country out of the wilderness.

Table 4.3  UPR and Microaggressions

| | Indirectly | Directly |
|---|---|---|
| Engaging the Concept of Microaggressions, and How It Applies to Your Communicated UPR | By being better able to accept client's reactions to microaggressions in and out of session | By helping you understand when you might have communicated conditional regard subtly or accidentally |

Do you have an increasing awareness about how privilege, power, and stereotypes operate beyond your awareness to cause microaggressions? Can you think of a microaggression you have experienced in the last week? Can you think of one in which you took part more actively?

## Social Justice Advocacy Competencies

In addition to the broad multicultural competencies, the American Counseling Association (2003) has also endorsed social justice advocacy competencies. *Social justice* refers to "a perspective in which a central role for counselors is to work toward increased equity, fairness, human rights and to work to eliminate injustice, oppression and human violation" (Multicultural Counseling and Social Justice Competencies, 2013, para. 1). It is a movement that not only has been growing in strength in recent years, but also one that was influential to the development of the person-centered approach as is explained in the next section. The advocacy competencies articulate two main ideas: empowerment and social action, which can be realized at the student/client, school/community, or public arena levels. *Empowerment* is working directly with clients to remove systemic barriers, while *social action* refers to acting on group, organizational, policy, or legislative levels (Multicultural Counseling and Social Justice Competencies, 2013, para. 3). Table 4.4 depicts the Advocacy Competencies showing the two main ideas of acting with or on behalf of three systemic levels (Toporek, Lewis, & Crethar, 2009).

The person-centered movement has also articulated many views on empowerment and social action from the 1920s to today (Cornelius-White, 2005; Klien, 2010; PCE, 2010; Proctor, 2002; Proctor et al., 2006; Rogers, 1977; Rogers & Cornelius-White, 2013). At a most basic level, the person-centered approach empowers the client through unleashing power within the person and using power with the person to remove obstacles inherent in the power difference between counselors and clients

Table 4.4 Social Justice Advocacy Competencies

| Client | Community | Public |
|---|---|---|
| Empower | Collaborate | Inform |
| Individually Advocate | Systemically Advocate | Politically Advocate |

*Source*: Toporek, R. L., Lewis, J. A., & Crethar, H. C. (2009). Promoting systemic change through the ACA Advocacy Competencies. *Journal of Counseling & Development*, 87, 260–268.

and between the client and the conditions that block them in the world through helping them become more open to their experience (Proctor, 2002; Rogers, 1977). At a more systemic level, Rice (2013) poignantly describes how approaches other than the nondirective person-centered approach disempower the person, beginning in childhood with a process that is replicated throughout other oppressions with adults. Hopkins (2013) and Schmid (2013) among others (Proctor et al., 2006) describe how the person-centered approach relates to social activism. They believe that person-centered practitioners will develop and become activists the more they practice and integrate an understanding of people different from them into their own experiences. The following section describes the history of multiculturalism, including social justice concerns, and how the person-centered approach influenced and was developed through its history.

# History of Multiculturalism and the Person-Centered Approach

Rogers's first published writing concerned his travels as a young man to East Asia in 1922. He engaged world leaders like the Chancellor of Germany, Supreme Court Justices in Korea, and the Chair of the World Student Christian Federation as well as thousands of people from all walks of life. It was this trip to several countries, interacting with people from around the world, that yielded ideas that would come to fruition across his lifelong writings. These themes included engaging inhumanity and oppression through nonviolence, openness to diverse experience, and learning through relational dialogue and interconnection. Each of these would be realized through practical applications in the person-centered approach (Rogers & Cornelius-White, 2013).

Early studies of the nondirective person-centered approach showed that it reduced prejudice and fascism in groups during and after World War II as compared to a traditional class format (Cornelius-White & Harbaugh, 2010). The 1960s and 1970s saw the development of the encounter group and civil rights movements in which the person-centered approach was a clear participant. In his late life during the late 1970s and 1980s, Carl Rogers negotiated peace in conflict zones like apartheid South Africa, war torn Northern Ireland, and in cold war "opponent" USSR. Rogers was a beacon of *cross-cultural communication* and a staunch opponent of *social injustice* and an *advocate* for children, women, people of color, labor, and anyone disenfranchised by the social situations in which they found themselves (Kirschenbaum, 2009). Indeed, Carl Rogers was nominated for the Nobel Peace Prize when he died. The person-centered cross-cultural workshops that began in that decade continue today. Finally, global research in the most recent decade with over 5,000 participants has shown how minority practitioners (people of color) have a greater preference for person-centered approaches than other major approaches to counseling, such as psychodynamic, cognitive behavioral, or systemic (Elliott, Orlinsky, Klein, Amer, & Partyka, 2003).

> **QUESTIONS TO LEARN WITH**
>
> - Have you taken a trip that or encountered a type of person who shifted your thinking?
> - What experiences changed your world? Hurt you? Challenged you? Lead you to faith or renunciation?

## Ethical and Religious Concepts and the Person-Centered Approach

The person-centered approach is the best-known humanistic model to counseling. As such, it is often considered secular, a practice that developed with intentional connections to science and lived experience rather than belief or dogma. Likewise, most diversity coverage focuses on the visible elements of race, gender, and disabilities, and tends to avoid issues related to religion and faith practices. However, encountering spiritual concepts and differences can be one of the most fundamentally altering and development-oriented means to improving one's fitness in multiculturalism (Cornelius-White, 2005).

Regarding ethics, the person-centered approach is a deliberate attempt to empower through a disciplined practice of respecting each person through noninterference beyond skilled companionship. It is an ethical practice where the means and ends are consistent and the self-determination (with a wide variety of ideas of what self might mean) of each client is a central tenet. In these respects, one might consider the person-centered approach to be somewhat a-religious. Nevertheless, three religious concepts from four traditions each intersect with the practice of person-centered therapy: the Taoist concept of *wu-wei*, the Christian concept of *grace*, and the interfaith and secular concepts of *nonattachment/mindfulness*.

## Wei-wu-wei

Freire (2009) has described how the concept of wu-wei relates to the person-centered approach. Wu-wei means *nondoing, nonaction,* or *without controlling*. Wei-wu-wei therefore means *action without action* or *effortless doing*. It is a form of releasing power within through harmony rather than pushing from outside. Lao-tzu in the Tao Te Ching (MacDonald, 1996) describes perhaps the best known description of wu-wei:

> That which offers no resistance,
> overcomes the hardest substances.
> That which offers no resistance
> can enter where there is no space.
> Few in the world can comprehend
> the teaching without words,
> or understand the value of non-action.[1]

The person-centered approach aims for a powerful, influential harmony without controlling others. The person-centered therapist understands and trusts the power of being in tune to help release a person's spirit and capacities. Likewise, the person-centered therapist recognizes that humility and patience help the client bravely face whatever they are going through, which is paradoxically the fastest way to help someone resolve their concerns or develop into a more functional person.

### QUESTION TO LEARN WITH

- You probably have a view of therapy or healing that involves therapist action, your expertise, or how persons need others to teach or rescue them somehow from their problems. How does wei-wu-wei sit with these ideas?

## Grace

The concept of *grace* in Christian theology is quite simply the love God grants to people without anyone having to earn it, incorporating mercy and forgiveness. One way to describe the person-centered therapist is to say that believing in every client's capacity and granting each one acceptance

---

[1] Tao #43, *Tao Te Ching*, Lao-tzu. A translation for the public domain by J. H. MacDonald, 1996.

and love regardless of circumstances or previous, current, or future actions model grace. The counselor's own experience of empathy shines that light of grace into the darkest, hardest to accept experiences a client presents. Likewise, grace can help explain a therapist's own experiences, forming the basis for self-acceptance.

> **QUESTION TO LEARN WITH**
>
> - What do you not accept about yourself? What elements of your past or present situation do you fear or would feel vulnerable to share with another person? With what issues would believing in grace challenge you?

## Nonattachment and Mindfulness

Though perhaps most identified with Buddhism, *nonattachment* is an important concept in Christianity, Hinduism, Jainism, Islam, Bahai, and secular healing practices (Detachment, n.d.). Stated most simply, nonattachment is freedom from desire and the self-referential connections to people and the world. It is *mindfulness* to the reality of things, a nonpossessive compassion and appreciation for all with whom we come into contact. In Buddhism, attachment is the source of suffering. Letting go of expectations and possessive attachments is the path to mindfulness and mindfulness the path to nonattachment. As long as one desires, then one is vulnerable to frustration and loss, but in freely living one moment to the next an enlightened state is possible (Rahula, 2007). In Christianity, the Latin phrase *incurvatus in se* (curved inward on oneself) conveys the central nature of how our identifications and pride are among the greatest sins. John the Apostle, Martin Luther, St. Ignatius, and others articulated this philosophy (Johnston, 2009).

The person-centered therapist aims to free herself from specific goals regarding her client both in the moment and enduring. She has only general goals of helping the client learn, grow, solve problems, protect agency, and generally become all the client can become. In practical terms, this means the person-centered therapist is as patient and accepting with desperation, fear, and self-negation as with success. Each emotion and experience of the client is viewed compassionately but dispassionately, revering it as an opportunity and something that can be faced to the extent the client faces it. In this way, the person-centered therapist may have a profound, peaceful, yet paradoxically powerful experience of helplessness to change the client in the moment. In the same vein, the person-centered therapist does not aim for the client to graduate from school, get married, have a child, succeed at work, or any such typically valued outcomes, but instead to find his or her path. There is a trust in the person-centered therapist that in being a witness and nonattached yet intimately close companion, the person will choose and develop prosocially but not with attachment to conventional outcomes. The phrase "trust the process" is popular among counselor educators to convey this faith in *being with* to facilitate rather than prescribing what a client should do or believe or choose.

### QUESTIONS TO LEARN WITH

- What beliefs do you hold? What do you identify with? With what issues might nonattachment be difficult to practice in encountering a person different from you?
- What kinds of people or experiences lead you to fear or desire? What combination of race, gender, sexual orientation, disability, or religious perspective challenges a freedom from expectations?

## The Paradox: A Universal System of Adaptability to Differences

The practice of person-centered therapy is remarkably monolithic and simplistic, namely to be honestly empathic and accepting, recognizing that each person is unique. Individuals are immersed in culture that is gendered, ethnic, sexual, ability oriented, age identified, economics based, and grounded in an infinite number of other group identifications. Furthermore, they have a unique autobiography of internal and external thought processes, feelings, and experiences. In this sense, person-centered therapy is a universal system that adapts to differences (Patterson, 1996/2000).

When a person-centered therapist aims to understand another, they often must hold the tension of all these bits of that person's context. It may include an adjustment to understand a person's entire family or culture or how their geography defines their identity. It will challenge the therapist herself to accept viewpoints and experiences she has never had as her own to further her client's journey. It includes that the therapist will need to understand her own culture, family, class, and a myriad of other parts of her identity. Each difference she encounters helps reveal a new part of herself or an old part that was never acknowledged or accepted.

What may have been perceived as simple becomes in practice exceedingly complicated. How do you empathize with potentially anyone?

How do you accept things that are dramatically different than your own beliefs or that you have never considered? How do you allow yourself to be open to being changed by each person, especially persons struggling to hang on for dear life, or who may be seen as disgusting or disturbed by many people? The practice of this book is to find answers to each of these questions. The purpose of this chapter is not only to introduce you to the role that multiculturalism and person-centered approach played in developing each movement, identify some ethical and religious concepts that can explain or enrich the practice, but it is also to invite you to explore your own challenges in practicing the person-centered approach in multicultural context.

> **QUESTIONS TO LEARN WITH**
>
> - How would you explain the multicultural paradox of the person-centered approach?
> - In what ways is the paradox not satisfying? How could you elaborate its shortcomings? What elements of multiculturalism do you need to explore beyond this chapter to understand how to better practice and appreciate multicultural person-centered therapy?

## Summary

- Multicultural competencies involve awareness of one's own culture and biases, understanding of clients' worldviews, and practicing culturally appropriate intervention strategies.
- Congruence, empathic understanding, and unconditional positive regard are universally applicable but culturally bound in their internal and communicated applications.
- Concepts like stereotypes, privilege, and microaggressions help foster culturally grounded core conditions and multicultural competencies.
- Person-centered approaches have a rich history related to social advocacy and the two movements may further intertwine and propel each other into the future.
- Ethical and religious concepts such as wei-wu-wei (Taoism), grace (Christianity), and nonattachment/mindfulness (Buddhism and Secularism) can help enrich the understanding and practice of person-centered approaches for counselors.

# 5

# A Case Illustration of Person-Centered Counseling

For you to have a better sense of what a contemporary person-centered approach looks like in action, consider the fictitious case of Deena and Casey. The case may allow you to see therapist (empathy, warmth, and genuineness) and client (contact, incongruence, and perception) conditions operating in a feedback loop in practice. It invites you to consider developments and multicultural elements and reflect on how the situation might relate to your life as a clinician.

## Deena

Deena is a 33-year-old woman who did not report her ethnicity but appears light skinned with dark features. She reported that she has problems with alcohol and depression. She also reported having just been released from an inpatient unit after attempting to commit suicide by overdosing on Tylenol. She reports living with her partner, Shauna, and Shauna's 12-year-old daughter. Deena wants help with her depression so that she can stay out of the hospital and not "freak out" Shauna and her daughter again. Deena reports having part-time work and earning less than $20,000 a year.

## Casey

Casey is a 25-year-old graduate student in counseling on her first internship. She completed practicum over the summer at a university-based training facility and is serving in an outpatient sliding scale clinic within a suburban setting. Casey is Caucasian and a first generation college student, who has done well in school and is passionate to help people. She has some concern about seeing someone on her first internship who recently attempted suicide but had worked with a client with suicidal thoughts on her practicum. Her onsite supervisor is very supportive, as is her university program supervisor, and she knows that she will need to frequently consult with them on Deena's case. The internship site uses the Session Rating Scale and Outcome Rating Scale (with permission from the authors of the scales) with all its clients.

## Session 1: First Encounter

Deena is dressed in jeans and a "Popeye" T-shirt (the T-shirt depicts the cartoon character Popeye eating spinach and about to rescue his beloved Olive Oyl). Deena willingly completes the Outcome Rating Scale at the start of the session, reporting a very low score of 4. (See Figure 5.1 at the end of this chapter for a graph of scores throughout her time with Casey.) The Outcome Rating Scale (ORS) asks the client to rate how well they are doing on 4 subscales (Individually, Interpersonally, Socially, and Overall) that are scored from 1 to 10, with a maximum total of 40. The Outcome Rating Scale is taken by the client at the beginning of the session. The Session Rating scale (SRS), which also comprises 4 subscales (Relationship, Goals & Topics, Approach & Method, and Overall), is taken near the end of the session; typically, the clinician will allow enough time at the end to briefly address the scores with the client if the score does not reflect near complete satisfaction.

Deena's score of 4 out of a possible 40 on the ORS indicates that Deena is struggling in all areas of her life. She averts eye contact with Casey and talks slowly. They begin by talking about her experience in the hospital.

Casey: Could you tell me a little more about how it felt to be there?

Deena: Desperate.

Casey: You felt desperate about being in the hospital.

Deena: Getting there. Ending up there. It was just total desperation. I mean, it was suicide . . .

Casey: You were so desperate that suicide seemed the only way out.

Deena: Exactly. (silence) I'm just totally bummed out. It was so humiliating to be there and having everybody ask me questions and all the while, I'm thinking it was all just a huge mistake.

Casey: You are ashamed.

Deena: The biggest mistake of my life.

Casey: And meanwhile, everyone is asking you questions.

Deena: "Why'd you do it? How were you feeling at the time?" Oh, just peachy. Wonderful. I mean, doesn't everybody overdose because they are just too damn happy?

Casey: They were asking you really obvious questions.

Deena: You'd think they'd just put down, "Patient's life completely sucks."

Casey: Sure.

Casey feels comfortable with Deena and "likes" her. It's fairly easy for her to follow Deena and feel compassion for her situation. Casey finds the discipline of the Nondirective attitude fairly simple with Deena because it's clear that Deena has much to say and is clearly in distress. With each empathic statement, Deena seems to open up more. While Deena continues to avoid eye contact with Casey, Deena's response of "Exactly" helps Casey feel confident that she is connecting and being useful to Deena. And when Deena responds with sarcastic humor ("Oh, just peachy. Wonderful") her mood seems to palpably lift.

Deena: Well, my life doesn't suck quite as much. But it still sucks.

Casey: Semi-sucky.

Deena: (laughs) Yeah, partly sucky with a chance of showers.

Casey: (smiling) I enjoy your sense of humor.

Deena: Better to laugh than cry. Shauna says she can tell when I'm getting bad when I do both at the same time.

Casey: Sometimes it seems hard to know which one to do, so you do both.

Deena: Yeah. (silence) There's no way in hell I'm putting them through that again.

Casey: Mhm, hmm.

Deena: I'm just not doing it. Just not.

For the remaining 10 minutes of the session, Casey carefully explores Deena's thoughts on suicide. Deena denies any intent, plan, or means to commit suicide. She reiterates that she won't put Shauna and her daughter through it again, and laughs, "I'm practically on lockdown as it is; someone is always watching me." Casey checks in with her internal experience, finds that she feels convinced by Deena's firm commitment to avoid thinking of suicide as an option, and is encouraged by Deena's ability to find humor in her situation. She tells Deena that, as she had explained in their first meeting, she will be checking in with her onsite supervisor and university supervisor so that she (Deena) can receive the best care possible.

Deena completes the Session Rating Scale at the end of the session (rating the session as highly satisfying) and as she is leaving, she pauses at the door and turns. She makes direct eye contact and says, "Thank you, Casey. Really."

## Session 2: Scared, Really Scared

Deena begins the second session silently. She speaks in short sentences with long pauses between, mostly about the stress in her relationship with Shauna. She appears sluggish, almost to the point of catatonia. During Deena's silences, Casey finds herself carrying on an internal dialogue with herself:

Deena: (silence)

Casey: (*I wonder if I should say anything?*)

Deena: (silence)

Casey: (*I know that I'm supposed to be comfortable with silence. What kind of silence is this? Awkward? Pregnant? Companionable?*)

Deena: (silence)

Casey: (*I wonder if I'm doing something wrong. Isn't she supposed to say something? I'm judging her. That's not nondirective, that feels judging. Maybe I should bring up her drinking?*)

Casey is taken off guard at how quiet Deena is at the beginning, especially given how forthright she was last time. She tries to be patient, but struggles a bit internally with keeping unconditional positive regard flowing and avoiding judging her client or herself. After a couple of minutes of silence, which feels like an eternity to Casey, she asks Deena about how her relationship with Shauna is going. Deena responds with more animation:

Deena: We fought a few days ago. Bad.

Casey: Ok.

Deena: She keeps on my case about my drinking.

Casey: What's that like for you?

Deena: Scared, very scared.

Casey: Mhm, hmm.

Deena: (silence)

Casey: (silence)

Deena: Scared, very scared.

Casey: You are so scared right now.

Deena: (silence) I don't disagree with her.

Casey: You agree with her concerns about your drinking?

Deena: It's too much. It's most days a six-pack, more on weekends or when I'm alone.

Casey: So you're drinking six beers most days, and more when the weekend comes or when you are all by yourself.

Deena: I don't disagree with her.

Casey: Mhm, hmm.

Deena: Scared, very scared.

Casey: Scared, very scared.

Deena: (silence)

Casey feels more comfortable during this spell when Deena is more expressive. Deena goes on to talk about what she likes to drink ("I like to think I'm a connoisseur of microbrews. I prefer dark beers like Imperial Stouts") and talks about the history of her drinking from early adolescence in relation to her mom's drinking ("Mom was one of those 50's housewives, you never would've known she drank, always vodka, secretive and at night"). When Deena again goes quiet, Casey feels calmer and surer that Deena was feeling understood, and that she is silent in order to gather her thoughts. This allows Casey to feel more unconditional and trusting of the process that Deena is "in" today. Nevertheless, Casey wonders about the strange repetition of "Scared, very scared," and whether her saying "Mhm, hmm" and "You are so scared right now" and then just "Scared, very scared" is helpful.

Casey: I'm wondering about "scared, very scared." Are these feelings of being scared related to feeling at risk for suicide? Are you feeling safe?

Deena: No.

Casey: No, you're not feeling safe?

Deena: No, I can't imagine doing anything like that now.

Casey: Do you have any feelings of wanting to hurt yourself?

Deena: No.

Casey: But you are feeling scared.

Deena: Exactly.

Casey again feels confident that Deena is not at present risk for suicide. She is concerned about the drinking, which can be a risk factor, and will bring this concern up with her supervisors. However, Deena has been consistent in her disavowal of entertaining suicide as an option. Casey's intuition is that Deena just needs someone to be with her in her feeling of being scared. Her intuition appears justified by Deena's remark, "Exactly" and by the SRS score, which remains at a nearly perfect level. The session time thus comes to an end. While Deena did become somewhat more voluble, she is still very quiet and not forthcoming with words or movement as she leaves.

## Session 3: Some Relief

Deena presents dressed in nice pants with a button-down plaid shirt tucked in. She appears to have had some sun on her face and neck. She speaks quietly, but quicker than in the previous session, talking about how her weekend went well. She reports that she "got an extra couple of shifts at work" and "actually enjoyed hanging out" with a few people after work. She discussed her relationship with Shauna more, explaining how she was "lost" before she met Shauna a couple of years ago, and about her mother who passed away about 6 months ago.

Casey is relieved that Deena appears more talkative and had a jump in her score on the Outcome Rating Scale, especially in the item related to social life, work, and friends, and somewhat on the Overall item. Casey had been curious about Deena's relationships with her mother and Shauna and was interested to learn about the recent loss of her mother. This session appears to provide them both with some relief—for Deena, that she has been able to get out and socialize, and for Casey, that Deena appears to be experiencing some positive changes in her living and working situations.

## Session 4: A Possible Misunderstanding

Deena begins by asking Casey what she thinks of her; Casey, in turn, struggles with staying present and not being drawn into an interior dialogue with herself:

Casey: You are really wondering about my thoughts about you.

Deena: Yes. What do you think of me?

Casey: (*What should I say? She's vulnerable. I certainly can't say You're a depressed, unpredictable mess! Oh, I smiled, I hope she doesn't misinterpret that.*)

Deena: I mean, I don't mean to be rude. I'm just being, I don't know, I just would like to know what you think of me. Whether it's what other people think of me.

Casey: You want to know what I think of you, whether it's similar to or different than what other people think of you. (*Ok, now I'm seeing where this is coming from. But it seems like she really wants me to tell her what I think. Will she get offended if I don't give her an answer? Should I answer her directly? What could I say?*)

Deena: I don't know, maybe it's just that I don't like myself and think you don't like me.

Casey: Mhm, hmm.

Deena: Do you know what's weird? There's this whole obsession with lesbians, like all the young girls always put that they are bi on their pages, their Facebook or Instagram or whatever, but then no one knows what to do with me standing in front of them. They just hate having to think. And end up hating me for making them do it.

Casey: There are these weird and conflicting things about people's responses to you as a lesbian. It makes you feel that they hate you, because they have to think about how to relate to a real human being who expresses her sexuality differently.

Deena: Yeah. It's kind of like me being depressed all the time. Like people are saying to me, "Get better." Well, what does getting better look like?

Casey: Mhm, hmm.

Deena: People can be so simplistic. It's like I'm smarter than most of them, but then I just fall back into feeling like I don't know anything.

Casey: You feel more on top of things in some ways, but also at a loss.

Deena: Yeah.

Casey was struck by the opening of the meeting, though she quickly found her bearings. Throughout the rest of the session, she continues to muse, *What do you do when a vulnerable client asks "What do you think of me?"* When Deena started to talk about why she asked, Casey found it easier to stay unconditional and her fear of judging or saying the wrong thing dissipated. She reflects that Deena seems to have some contradictions today—pride coupled with doubt in several areas.

At the end of the session, Deena takes a minute to fill out the Session Rating Scale. She marks the items on the SRS concerned with "Approach or Method" and "Goals and Topics" at the midpoint, indicating that her satisfaction with the session has dropped.

When Casey takes the SRS from Deena, she notices the big drop in scores. She knows she is supposed to address any score beneath 36 directly in session and had practiced doing this with the other students, but now that the moment of truth has arrived, she feels unsure about how to do it.

She decides that "less is more" and paraphrases the ORS items that Deena had marked low:

Casey: I noticed we did not talk about what you wanted and that my approach was not as good a fit today.

Deena's response is not clear to Casey. Deena says a few different things, which Casey has trouble following. The internal dialogue that Casey struggles with is drowning out what Deena is saying: the only things that she remembers is that Deena said something about wishing Casey would be more "active" and give more "guidance." Casey feels a little defensive, but tries her best to take the feedback.

Casey: Ok, so you wish I would have given you more advice today?

Deena: Maybe. I know you can't tell me how to live my life. I just didn't feel as comfortable today.

Casey resolves that she might ask Deena more frequently if she is getting what she wants, especially when Deena asks her questions or presents more contradictions in the future. Casey also thinks she better talk with her supervisors and watch part of this session with them.

## Session 5: Nonstop Talker

Deena sits down and starts to talk about her trip to the Counseling Center.

Deena: So when that crazy lady cuts me off while I'm driving, I kept saying, "You fucking bitch! You fucking bitch!" Then I notice other drivers looking at me, and I had to laugh at myself.

Casey: You amused yourself.

Deena: Yeah, it was too comical, me going all road-ragey, and on my way to see my counselor so I can get all Zen about my life. (laughs)

Casey: A Zen laugh?

Deena: (still laughing) Yeah. I know, I sound like Woody the Woodpecker.

Casey: (laughing, too) It's a very infectious one.

Deena chatters on, describing what she ate for lunch, what she did over the weekend, and several things about work and home life with Shauna and her daughter. When Casey mentions that their time is about up, a look crosses Deena's face, a look that Casey has some difficulty deciphering.

Casey: The time has really flown.

Deena: Yeah.

Casey: You seem, I don't know, maybe a little surprised?

Deena: Yeah.

Casey: Was that it?

Deena: (sighs)

Casey: That's a big sigh.

Deena: I don't know, I just felt a little less . . . I guess it's just nice to be with someone.

Casey: It's nice to spend time together.

Deena: (standing) Well, Doc, our time is up.

Casey feels relieved that Deena is not asking her for advice and enjoys the lighter side of her personality, especially appreciating the cussing and smiling. She appreciates learning about the details of her life because sometimes things with Deena seem so emotional that she doesn't really know what Deena's life looks like. Casey is also relieved because the client before Deena seemed dramatic and difficult to deal with and Casey was a bit worried about what Deena would present today. The session felt to Casey like Deena was sad, and probably lonely, but that these emotions weren't as extreme as they were in other sessions.

## Session 6: I Miss My Mom, Bad. She Was a Bitch to Deal With

Deena missed her appointment one week after Session 5, but called the next day to reschedule. When she comes into Session 6, she is dressed in baggy shorts and a white T-shirt.

Deena: So I've told you about my boss, right?

Casey: I think so. She's, how did you put it?

Deena: (laughs) I probably called her a "ball-busting micromanager."

Casey: (smiles) That was it!

Deena: I just don't get why she doesn't see that she's creating problems for herself. She gives me something to do, looks over my shoulder, then when I hand it in to her, she redoes it. It's three times the work.

Casey: It sounds like she doesn't trust others to do their work, so she has to look over their shoulders.

Deena: It's so self-defeating.

Casey: Mhm, hmm.

Deena: And I know self-defeating. (smiles)

Casey: You recognize how your boss is creating problems for herself.

Deena: Yeah. (silence) This is totally random, but I wish I still played softball.

Casey: You played softball?

Deena: Third base. Batted cleanup.

Casey: In high school?

Deena: Yeah. Tried out in college but wasn't good enough. I only lasted less than a year, so it wouldn't have mattered anyway.

Casey: Was there a link to your workplace? Does your workplace have a softball team?

Deena: (startled) That's too weird.

Casey: What's that?

Deena: That's what I read in the local paper. There's a co-ed slow pitch league that was putting out a call for teams.

Casey: So they're looking for teams.

Deena: I used to love playing. We'd play all the time in my neighborhood. There were these three Mexican sisters that I used to play with. Consuela, Rosa . . . what was the third sister's name? Man. Something hard to pronounce . . . Lourdes!

Casey: That's a lovely name.

Deena: They moved away in high school. I remember being really bummed.

Casey: You were close to them.

Deena: They were so lively. They'd come over sometimes.

After reminiscing about the sisters, Deena begins to talk about her mother, saying how she remembers her mother coming to her games in high school. As she talks, her face looks sad. She explores some of her feelings for her mother and how life is different now that she is not alive.

Casey is following along with Deena just fine. When she switches from describing about her difficult boss, to reminiscing about baseball, and then missing her mother, Casey wonders whether she could be more "active" like Deena seemed to want a few sessions ago. After following her closely for a while, feeling confident that Deena was feeling understood, she decides to try an experiment.

Casey: Would you like to try something that might help resolve things with your mother's passing?

Deena: Umm. Sure.

Casey: So could you imagine your mother for me now. (pauses)

Deena: OK.

Casey: What she looks like. (pauses) The sound of her voice. (pauses) What . . . what she might be wearing on a warm day like this.

Casey pulls up a chair and positions it in front of them.

Casey: Can you imagine your mother sitting here now?

Deena: Yeah.

Casey: If she was here, what would you say to her?

Deena: I don't know.

Casey: Take your time.

They sit in silence for a couple of minutes. Then Deena speaks.

Deena: Maybe "I miss you"?

Casey: Uh-huh.

There is more silence.

Deena: Actually I think I would really want to say, "You stupid bitch. Why did you go and fucking die?"

Deena wells up and large tears roll down her cheek. Casey follows along, noticing the feelings and meanings Deena was expressing in a style similar to what she had done most frequently throughout their sessions, but here about Deena's dialogue with her "mom." At one point, Casey says, "How would your mom have responded to all that?" and Deena is able to respond. After about 6 minutes with Deena and her mother "talking" and Casey closely, sensitively following, Deena was quiet for about one minute. Then she said, "Wow. I didn't expect that." Deena gave her typical answers on the Session Rating Scale—very satisfied—and confirmed their time for next week.

## Session 7: Talkative Again

Deena came in the next week and again seemed light and talkative and sarcastic about the "crazy bitches" in her life, like her boss and Shauna's "mouthy daughter." She discussed a lot of events that happened during the week, like Shauna's mother's drama with Shauna's brother who is on disability from a stint overseas in the military. She also talked about a "girl at work" she's been hanging with more. She mentions briefly that she went to a baseball game with Shauna and that she sorted through and then "gave away, trashed, or kept some stuff from boxes of old junk" from her mother. She also talked about how she was signing people up at her workplace for the slow pitch league. She then went quickly back to talk about more things happening in her life.

Casey again enjoyed the biting sense of humor that Deena sometimes displayed. She noticed that Deena was scoring most of the items on the Outcome Rating Scale above the halfway point for the first time. She was careful to respond slowly and thoughtfully when Deena spoke of sorting through her mom's boxes and about going to the baseball game with Shauna, even though Deena seemed to barrel right on after just a moment's pause. She felt good that Deena didn't completely ignore the strands that they had talked about before, but was glad to see that she was living and "inside" her own life more, paying attention to the details and events.

## Sessions 8 and 9: No Show

Casey was wondering how things were going for Deena when she didn't show for her eighth session. When she didn't call and didn't come the next week, she realized Deena might not be coming back. She reflected on her time with Deena and wondered "How did I do?"

As she reflected, she thought of the fact that Deena was displaying more of her humor. Deena's humor appears to be a very significant adaptive strategy, and she appeared to be able to use it frequently to help regain her balance. Casey also reflected on the success of their work together—the times when Deena responded with, "Exactly," and Casey felt that Deena was truly experiencing her empathy. She thought of the role play they did and Deena's ability to respond in her mother's voice. Deena clearly felt that this was powerful, and it seemed to be intensely therapeutic. In addition, the fact that Deena sorted through stuff from her mother seems highly significant to Casey, as if it is a concrete representation of Deena sorting through the emotional legacy from her mother. Casey thought about how Deena's mood seemed to be lifting, and that she was more lively and talkative as sessions wore on. She thought about the positive changes Deena seemed to be making—taking Shauna's daughter to a ball game, making a new friend at work, and signing people up for softball. Finally, she reflected that the ORS itself was the best and most reliable indicator that Deena was getting better. As we have discussed, the most reliable indicator of client progress comes from clients themselves, and thus Casey's reflection, while important, must take a backseat to Deena's own estimation of her progress. She also had some remaining worries—the drinking, which Casey doesn't have a great handle on, whether Deena is still doing it, and the fights with Shauna. Moreover, while Casey wasn't pleased at the abrupt ending of treatment, it served as yet another challenge to trust in her client's tendency to self-actualize or a failure of sorts.

## Questions for Your Consideration

- How do you think Casey did? What different sources can you use to help you answer that question?
- What are your impressions about Casey's reflections and worries? What would you have been concerned about? How confident would you have felt as an intern in these situations or after the treatment was over?
- At what points in the case presentation did you feel nervous? How would you have handled some of the things that happened?

- What multicultural issues were present, and how were they dealt with or not dealt with?
- Did you notice core conditions present? Where did they seem more or less obvious?
- What styles of person-centered counseling did you notice?

# Summary

- The case shows how empathy, congruence, and unconditional positive regard are experienced inside the counselor and in statements and behavior.
- It shows how contact can be fragile even if primarily enduring and how client incongruence, vulnerability, and perception fuels client expression and change.
- The dialogue and descriptions show how client perception, both in the moment and as measured by scales, can affect the counselor to adjust empathy to fit with the client.
- The case raises multicultural issues and shows some elements of different developments of person-centered approaches in action.
- The chapter invites the reader to reflect along with the counselor on what can be learned from the experiences involved.
- It shows how candor, grit, and human connection shine through.

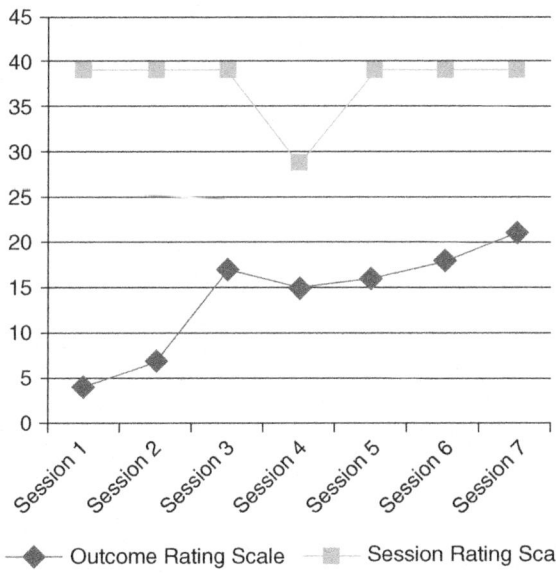

**Figure 5.1** Deena's Weekly Outcome Rating (ORS)* and Session Rating Scores (SRS)**

*Scores on the ORS beneath 25 represent a typical "clinical" presentation.

**Scores beneath 36 on the SRS signal a potential problem in the therapeutic alliance.

# 6

# Conclusion

Client, Relationship, Developments, and Multiculturalism: These are the main themes of *Person-Centered Approaches for Counselors* and in some respects the themes of the entire common-factors-based *Theories for Counselors* series. Like any of the theories explored in the series, person-centered approaches (PCA) are useful to the extent that they help value and foster the heroic qualities of clients, particularly the accordance of the counselor's approach with the client's perspective. The aim is to further the usefulness of the therapeutic encounter. This chapter revisits some of the ideas and applications discussed earlier and suggests new implications and challenges for you as you approach other theories and practices, beginning with the six conditions. Chapter one discussed three core "therapist relational" conditions—genuineness, empathy, and unconditional positive regard—while Chapter 2 discussed the three "client" elements—contact, incongruence, and perception.

## PCA Is Love

God is love; dog is love; the PCA is love. Ok, silliness aside, a main theme of common factors is relationship, and the operative element in person-centered relationships is *unconditional positive regard*, also known as nonpossessive love. *Empathy* is the means to deliver this love, by tailoring each therapist response to the culturally and individually specific experience of each client. Likewise, the love needs to be sincere, real, and delivered by a therapist who is self-accepting and congruent enough to deliver it with credibility. The three core conditions—empathy, unconditional positive regard, and *genuineness*—form the backbone of person-centered practice.

## Courageous, Compassionate Confrontation

The person-centered approach aims to help clients courageously confront their own experience to make meaning from confusion and ambiguity, change their perspective and actions, and become more open to experience the world anew. Clients need to have sufficient *contact* with their counselor, reality, and themselves to start this journey. The confrontation within the clients is with themselves, the *incongruence* or vulnerability they are experiencing.

## Perception

Client *perception* is the sixth condition and most vital part of the feedback loop that drives the therapeutic process, both from a common factors and person-centered approach. It is not enough that you experience self-acceptance, empathy, and unconditional positive regard toward your client. Clients need to perceive these conditions and then in turn make responses to help refine your connection to them and alter the display of the conditions to fit each person's unique cultural and idiosyncratic situation. The target of a person-centered counselor's statements is the universal needs for acceptance and understanding, needs that shift in presentation with each cultural and individual difference and also as each new meaning and feeling unfolds.

## Linking the Relational and the Client Conditions

It takes discipline on the part of the therapist to "be with" a client, not exert power over them, courageously shining unconditional light into each cranny with empathy. Pivotal to being with them is listening for their feedback. The client needs to perceive the therapist's loving communication and provide feedback to the therapist to adjust the light where it needs to go as each new feeling or meaning opens another window to unfolding experience. Each empathic statement is an invitation for the client to confront their experience and correct the accuracy of the counselor to what they are experiencing. Clients change their perceptions constantly so that an empathic statement that was spot on may be off once the client reflects on it and a new meaning or feeling unfolds. A sign of growth is that a person who had been angry for a while notices hurt, or another who was sad for a while notices how she is angry. In this way, clients can integrate different emotions and the wealth of memories, information, and motivation that accompany them. Likewise, whether it be through identifications of stereotypes, privilege, or microaggressions, the counselor is bravely confronting their own experience, too. Being with another means being open to learning and change in oneself alongside clients.

## It Figures

Figure 6.1 provides a simplified but hopefully useful way to conceptualize the six conditions with pictures and captions. The curative element in person-centered therapy is genuine unconditional positive regard (UPR). UPR helps the clients accept aspects of themselves and their experience, which they have been unable to on their own (Bozarth, 1998). Empathy focuses UPR to where it needs to go, connecting the therapist's

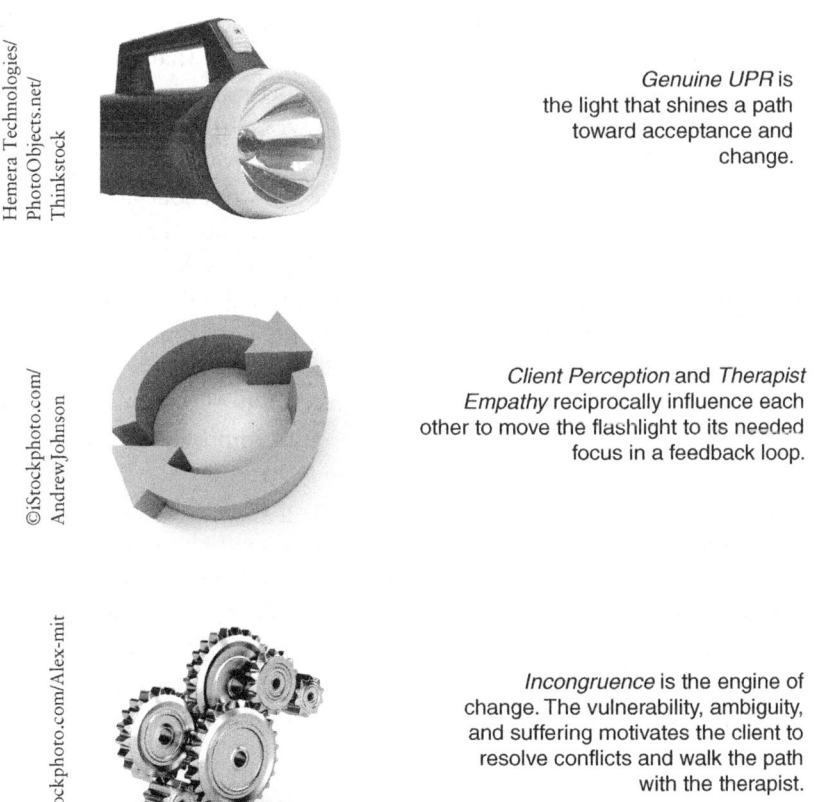

Figure 6.1   Heuristic Representations of the Person-Centered Conditions

understanding with the unique aspects of a client's experience. Clients perceive the UPR and help correct the therapist's empathy to their unique cultural and individual worldview as well as each moment's shifting experience, forming a feedback loop. Contact is connective precondition, linking clients with reality, themselves, and the therapist.

## Understand, Verify You Understand, and Improve

"Understand. Just understand." These were the words my mentor told me when I told her I was so nervous going into my first therapy sessions. At a most fundamental level, the primary action of the person-centered therapist is empathic statements. Client perceptions and therapist empathy form an ever-present moment-to-moment feedback loop. But how do you know how successful your empathic statements were in furthering the self-exploration, clarification, and motivational processes of the client? How accurately can you receive the client's perceptions of your empathic communications? Also, since we know from research that therapists are worse judges of the strength of their genuine, unconditional empathy than observers (such as supervisors) and especially worse than clients, the most accurate judges (Norcross, 2002), how can you improve your judgment?

The common factors approach of Duncan (2013) and colleagues provides one sure fire way, supported by strong research: Give the client simple, quick assessments at the start and end of each session. They have validated the Session Rating Scale and Outcome Rating Scale, which take under 1 minute to administer to clients. You can download and use copies from https://heartandsoulofchange.com/content/measures/login.php. The Session Rating Scale provides immediate feedback to the counselor at the end of the session to know whether to broach the subject of how to improve your services or whether the client is highly satisfied and likely to improve. The Outcome Rating Scale provides information at the start of the session about the client's functioning during the previous week so that counselors can tell quickly whether improvements expected (such as some movement in the first 3 or 4 weeks) are occurring. If improvement is not occurring, this again provides a vehicle to broach the subject about whether your services are adequate or a change is needed. In this way, you can have an objective corrective device to help improve your assessment and communication of the core conditions.

## Power Within, Power With, Power Over

PCA has deep social justice roots. It empowers the client to release their power within, the actualizing tendency motivated through incongruence to become more open to and accepting of their full experience and potential. It accomplishes this by sharing power, relinquishing the structural power differentials of cultural variables (e.g., privilege, status, prejudice) and role power of the therapist being the less vulnerable party in the interchange. Despite its identification as an approach to psychotherapy, the PCA developed out of large cross-cultural and educational dialogues (Cornelius-White & Harbaugh, 2010; Rogers & Cornelius-White, 2013) into a wide variety of interdisciplinary and political activities (Cornelius-White, Motschnig-Pitrik, & Lux, 2013a; Kirschenbaum, 2008). PCA preserves the gift of clients not only *having a solution* to their problems but also *solving* their problems.

## The Ever-Reaching Actualization Process: Give and Receive

The person-centered approach is about helping people maintain and enhance. As a counselor, you are on a never-ending journey to improve your services. On that journey, you will learn about yourself. Opening yourself to actively listen to another is a transformative process. You need to experience the world through another's eyes, which inevitably changes the view from your own eyes. Perhaps to a client, as important to receiving honest empathy and unconditional positive regard in therapy, is learning to build relationships in your daily life based on the modeled behaviors of your counselor accepting and understanding themselves and the client. Clearly, the reciprocity and mutuality of relationships is one of the most satisfying and gratitude producing there is. To receive is good, but perhaps giving is even better.

## Summary

- Nonpossessive love is the working element in person-centered approaches for counselors.
- Through collaborative, courageous, compassionate confrontation of the client's difficult and conflicting experiences, clients develop.
- Client perception of a therapist's empathy, warmth, and genuineness creates a feedback loop for continual understanding and improvement by client and therapist. This can be a moment-to-moment adjustment and a measured feedback through assessments like the Session Rating Scale and Outcome Rating Scale.
- Genuine UPR can be thought of as a light that is focused where it needs to go by therapist empathy and client perception, which allows the engine of a client's incongruence to move them in keeping with their actualizing tendency.

# References

Ahn, H., & Wampold, B. E. (2001). Where oh where are the specific ingredients? A meta-analysis of component studies in counseling and psychotherapy. *Journal of Counseling Psychology, 48*, 251–257. doi: 10.1037//O022-OI67.48.3.251

American Counseling Association. (2003). Advocacy competencies. Retrieved from http://www.counseling.org/Resources/Competencies/Advocacy_Competencies.pdf

Ampersand. (2006). A list of privilege lists. Retrieved from http://amptoons.com/blog/2006/09/26/a-list-of-privilege-lists

Association for Multicultural Counseling and Development. (1992). Cross-cultural competencies and objectives. *Journal of the Association for Multicultural Counseling and Development*. Retrieved from http://counseling.org/docs/competencies/cross-cultural_competencies_and_objectives.pdf

Baldwin, M. (2000). Interview with Carl Rogers on the use of self in therapy. In M. Baldwin (Ed.), *The use of self in therapy* (pp. 29–38). New York, NY: Haworth.

Barrett-Lennard, G. (2013). Relationship worlds and the plural self. In J. Cornelius-White, R. Motschnig-Pitrik, & M. Lux (Eds.), *Interdisciplinary handbook of the person-centered approach: Research and theory* (pp. 277–288). New York, NY: Springer.

Bike, D. H., Norcross, J. C., & Schatz, D. M. (2009). Processes and outcomes of psychotherapists' personal therapy: Replication and extension 20 years later. *Psychotherapy, 46*, 19–31.

Bohankske, R. T., & Franczak, M. (2010). Transforming public behavioral health care: A case example of consumer-directed services, recovery, and the common factors. In B. L. Duncan, S. D. Miller, B. E. Wampold, & M. A. Hubble (Eds.), *The heart and soul of change: Delivering what works* (2nd ed., pp. 299–322). Washington, DC: American Psychological Association.

Bohart, A., & Tallman, K. (1999). *How clients make therapy work: The process of active self-healing*. Washington, DC: American Psychological Association.

Bohart, A., & Tallman, K. (2010). Clients: The neglected common factor. In B. L. Duncan, S. D. Miller, B. E. Wampold, & M. A. Hubble (Eds.), *The heart and soul of change: Delivering what works* (2nd ed., pp. 83–112). Washington, DC: American Psychological Association.

Bozarth, J. (1993). Not necessarily necessary but always sufficient. In D. Brazier (Ed.), *Beyond Carl Rogers: Towards a psychotherapy for the 21st century* (pp. 92–105). London, United Kingdom: Constable.

Bozarth, J. D. (1998). *The person-centered approach: A revolutionary paradigm.* Ross-on-Wye, United Kingdom: PCCS Books.

Brodley, B. T. (2005). Introduction: About the non-directive attitude. In B. Levitt (Ed.), *Embracing nondirectivity: Reassessing person-centered theory and practice in the 21st century* (pp. 1–4). Ross-on-Wye, United Kingdom: PCCS Books.

Brodley, B. T., & Lietaer, G. (2006). Rogers' transcripts, *Vol. 5*. Retrieved from http://www.diretoriopsicologos.com/terapeutas

Brown, A. P., Marquis, A., & Guiffrida, D. A. (2013). Mindfulness-based interventions in counseling. *Journal of Counseling and Development, 91*, 96–104.

Constantine, M. (2007). Racial microaggressions against African American clients in cross-racial counseling relationships. *Journal of Counseling Psychology, 54*, 1–16.

Constantine, M., & Sue, D. (2007). Perceptions of racial microaggressions among black supervisees in cross-racial dyads. *Journal of Counseling Psychology, 54*, 142–153.

Conyers, L. M. (2002). Disability: An emerging topic in multicultural counseling. In J. Trusty, E. J. Looby, & D. S. Sandhu (Eds.), *Multicultural counseling: Context, theory and practice, and competence* (pp. 173–202). Huntington, NY: Nova Science.

Cook, J. M., Biyanova, T., Coyne, J. C. (2009). Influential psychotherapy figures, authors, and books: An internet survey of over 2,000 psychotherapists. *Psychotherapy: Theory, Research, Practice, Training, 46*, 42–51.

Cooper, M. (2008). The facts are friendly. *Therapy Today, 19*(7). Retrieved from http://www.therapytoday.net/article/show/212

Cornelius-White, J. H. D. (2002). The phoenix of empirically supported therapy relationships: The overlooked person-centered basis. *Psychotherapy: Theory, Research, Practice, Training, 39*, 219–222.

Cornelius-White, J. H. D. (2003). Person-centered multicultural counseling: Rebutted critiques and revisited goals. *Person-Centered Practice, 11*(1), 3–11.

Cornelius-White, J. H. D. (2005). Teaching person-centered multicultural counseling: Experiential transcendence of resistance to increase awareness. *Journal of Humanistic Counseling Education and Development, 44*, 225–240.

Cornelius-White, J. H. D. (2006, May). Cultural congruence: Subtle veils of whiteness and patriarchy. *Person-Centered Quarterly*, 4–6.

Cornelius-White, J. H. D. (2007a). Congruence as extensionality. *Person-Centered and Experiential Psychotherapies, 6*, 196–204.

Cornelius-White, J. H. D. (2007b). Congruence: An integrated five dimension model. *Person-Centered and Experiential Psychotherapies, 6*, 228–238.

Cornelius-White, J. H. D. (2007c). Environmental responsibility: A social justice mandate for counseling. *Journal of Border Educational Research, 6*, 2, 5–16.

Cornelius-White, J. H. D. (2007d). Learner-centered teacher-student relationships are effective: A meta-analysis. *Review of Educational Research, 77*, 113–143.

Cornelius-White, J. H. D. (Ed.). (2012). *Carl Rogers: The China diary.* Ross-on-Wye, United Kingdom: PCCS Books.

Cornelius-White, J. H. D. (2013). Congruence. In M. Cooper, M. O'Hara, P. F. Schmid, & A. Bohart (Eds.), *The handbook of person-centered psychotherapy and counseling* (2nd ed., pp. 168–181). New York, NY: Palgrave Macmillan.

Cornelius-White, J. H. D., & Godfrey, P. (2004). Pedagogical crossroads: Integrating feminist critical pedagogies and the person-centered approach in education. *Encountering feminism: Intersections of Feminism and the Person-Centered Approach*. Ross-on-Wye, United Kingdom: PCCS Books.

Cornelius-White, J. H. D., & Harbaugh, A. P. (2010). *Learner-centered instruction: Building relationships for student success*. Thousand Oaks, CA: Sage.

Cornelius-White, J. H. D., & Motschnig-Pitrik, R. (2010). Effectiveness beyond psychotherapy: The person-centered, experiential paradigm in education, parenting and management. In M. Cooper, J. Watson, & D. Holldampf (Eds.), *Effectiveness of the person-centered and experiential paradigm: Report of the World Association for Person-Centered and Experiential Psychotherapy and Counseling Taskforce*. Ross-on-Wye, United Kingdom: PCCS Books.

Cornelius-White, J. H. D., Motschnig-Pitrik, R., & Lux, M. (Eds.). (2013a). *Interdisciplinary applications of the person-centered approach*. New York, NY: Springer.

Cornelius-White, J. H. D., Motschnig-Pitrik, R., & Lux, M. (Eds.). (2013b). *Interdisciplinary handbook of the person-centered approach: Research and theory*. New York, NY: Springer.

D'Andrea, M., & Daniels, J. (2001). RESPECTFUL counseling: An integrative multidimensional model for counselors. In D. B. Pope-Davis & H. L. K. Coleman (Eds.), *The intersection of race, class, and gender in multicultural counseling* (pp. 417–466). Thousand Oaks, CA: Sage.

Davis, D. M., & Hayes, J. A. (2011). What are the benefits of mindfulness? A practice review of psychotherapy-related research. *Psychotherapy, 48*, 198–208.

Detachment. (n.d.). In Wikipedia. Retrieved from http://en.wikipedia.org/wiki/Detachment_(philosophy)

Duncan, B. L. (2002). The legacy of Saul Rosenzweig: The profundity of the Dodo bird. *Journal of Psychotherapy Integration, 12*, 32–57. doi: 10.1037//1053-0479.12.1.32

Duncan, B. L. (2013). The heart and soul of change: Getting better at what we do. *Iowa Psychologist, Summer*, 4–5. Retreived from https://heartandsoulofchange.com

Duncan, B. L., & Miller, S. (2000). *The heroic client*. San Francisco, CA: Jossey-Bass.

Duncan, B. L., Miller, S. D., & Hubble, M. A. (1999). *The heart and soul of change: What works in therapy*. Washington, DC: American Psychological Association.

Duncan, B. L., Miller, S. D., Wampold, B. E., & Hubble, M. A. (Eds.). (2009). *The heart and soul of change: Delivering what works in therapy* (2nd ed.). Washington, DC: American Psychological Association.

Duncan, B. L., Miller, S. D., & Sparks, J. (2007). Common factors and the uncommon heroism of youth. *Psychotherapy in Australia, 13*(2), 34–43.

Elliott, R., Orlinsky, D., Klein, M. J., Amer, M., & Partyka, R. (2003). Professional characteristics of humanistic therapists: Analyses of the Collaborative Research Network sample. *Person-Centered and Experiential Psychotherapies, 3*, 188–203.

Fisher, L. (2013). How can I trust you? Encounters with Carl Rogers and game theory. In J. Cornelius-White, R. Motschnig-Pitrik, & M. Lux (Eds.), *Interdisciplinary handbook of the person-centered approach: Research and theory* (pp. 299–318). New York, NY: Springer.

Focusing. (2014). An introduction to focusing: Six steps. Retrieved from http://www.focusing.org/gendlin/docs/gol_2234.html

Freire, E. (2009). A quiet revolution . . . or swimming against the tide? *Person-Centered and Experiential Psychotherapies, 8*, 224–232.

Gay, P. (2006). *Freud: A life for our time*. New York, NY: Norton.

Gendlin, E. (1982). *Focusing* (2nd ed.). New York, NY: Bantam.

Gendlin, E. (2014). Short form. Retrieved from http://www.focusing.org/short_gendlin.html

Grafanaki, S. (2001). What counseling research has taught us about the concept of congruence: Main discoveries and unresolved issues. In G. Wyatt (Ed.), *Congruence (Rogers' therapeutic conditions: Evolution, theory, and practice)* (Vol. 1, pp. 18–35). Ross-on-Wye, United Kingdom: PCCS Books.

Grant, B. (1985). The moral nature of psychotherapy. *Counseling and Values, 29*, 141–150.

Grant, B. (1990). Principled and instrumental nondirectiveness in person-centered and client-centered therapy. *Person-Centered Review, 5*, 77–88.

Greenberg, L. (2010a). *Emotion-focused therapy (Theories of psychotherapy)*. Washington, DC: American Psychological Association.

Greenberg, L. (2010b). Emotion-focused therapy: A clinical synthesis. *FOCUS, 8*, 32–42. Retrieved from http://focus.psychiatryonline.org/article.aspx?articleID=53063

Hays, P. A. (2008). *Addressing cultural complexities in practice: Assessment, diagnosis, and therapy* (2nd ed.). Washington, DC: American Psychological Association.

Hopkins, R. (2013). Staying human: Experiences of a therapist and political activist. In J. H. D. Cornelius-White, R. Mostnig-Pitrik, & M. Lux (Eds.), *Interdisciplinary applications of the person-centered approach* (pp. 229–234). New York, NY: Springer.

Hubble, M. A., Duncan, B. L., Miller, S. D., & Wampold, B. E. (2010). Introduction. In B. L. Duncan, S. D. Miller, B. E. Wampold, & M. A. Hubble (Eds.), *The heart and soul of change: Delivering what works in therapy* (pp. 23–46). Washington, DC: American Psychological Association.

Johnson, A. G. (2006). *Privilege, power, and difference* (2nd ed.). Boston, MA: McGraw Hill.

Johnston, M. (2009). *Saving God*. Princeton, NJ: Princeton University Press.

Keys, S. (Ed.). (2003). *Idiosyncratic person-centered therapy: From the personal to the universal*. Ross-on-Wye, United Kingdom: PCCS Books.

Killermann, S. (2013). Privilege Lists, AKA the Dirty 30 Lists. Retrieved from http://itspronouncedmetrosexual.com/category/privilege-lists

Kirschenbaum, H. (2008). *Life and work of Carl Rogers*. Washington, DC: American Counseling Association.

Kirschenbaum, H. (2009). *The life and work of Carl Rogers*. Alexandria, VA: American Counseling Association.

Kirschenbaum, H., & Henderson, V. (Eds.). (1989). *The Carl Rogers reader*. Boston, MA: Houghton Mifflin.

Klien, G. H. (2010). *Loss: A personal journey of empowerment*. La Jolla, CA: Person-Centered Press.

Kriz, J. (2013). Person-centered approach and systems theory. In J. Cornelius-White, R. Motschnig-Pitrik, & M. Lux (Eds.), *Interdisciplinary handbook of the person-centered approach: Research and theory* (pp. 261–276). New York, NY: Springer.

Lago, C. (2012). The person-centered approach and its capacity to enhance constructive international communication. In J. H. D. Cornelius-White, M. Lux, & R. Motschnig-Pitrik (Eds.), *Interdisciplinary applications of the person-centered approach* (pp. 201–212). New York, NY: Springer.

Lambert, M. J. (1992). Psychotherapy outcome research: Implications for integrative and eclectic therapists. In J. C. Norcross & M. R. Goldfried (Eds.), *Handbook of psychotherapy integration* (pp. 94–129). New York, NY: Basic Books.

Lambert, M. J. (2013). Enhancing psychotherapy through feedback to clinicians. Retrieved from http://www.e-psychologist.org/index.iml?mdl=exam/show_article.mdl&Material_ID=3

Lambert, M. J., & Barley, D. E. (2001). Research summary on the therapeutic relationship and psychotherapy outcome. *Psychotherapy, 38*, 357–361. doi:10.1037/0033-3204.38.4.357

Lambert, M. J., & Ogles, B. M. (2004). The efficacy and effectiveness of psychotherapy. In M. J. Lambert (Ed.), *Bergin and Garfield's handbook of psychotherapy and behavior change* (5th ed., pp. 139–193). New York, NY: Wiley.

Langton, L., Planty, M., & Sandholtz, N. (2013). *Hate crime victimization, 2003–2011*. Washington, DC: Bureau of Justice Statistics. Retrieved from http://www.bjs.gov/index.cfm?ty=pbdetail&iid=4614

Lipsitz, G. (2005). The possessive investment in whiteness. In P. S. Rothenberg (Ed.), *White privilege: Essential readings on the other side of racism* (2nd ed., pp. 67–90). New York, NY: Worth.

Luborsky, L., Singer, B., & Luborsky, L. (1975). Comparative studies of psychotherapy: Is it true that "everyone has won and all must have prizes"? *Archives of General Psychiatry, 32*, 995–1008.

Lux, M. (2013). The circle of contact: A neuroscience view on the formation of relationships. In J. Cornelius-White, R. Motschnig-Pitrik, & M. Lux (Eds.), *Interdisciplinary handbook of the person-centered approach: Research and theory* (pp. 79–94). New York, NY: Springer.

MacDonald, J. H. (1996). *Tao Te Ching*, complete online text translation for the public domain. Retrieved from http://www.wright-house.com/religions/taoism/tao-te-ching.html

Martin, D. G. (2010). *Counseling and therapy skills* (3rd ed.). Long Grove, IL: Waveland.

McIntosh, P. (2013). White privilege: Unpacking the invisible knapsack. Retrieved from http://www.amptoons.com/blog/files/mcintosh.html

Mearns, D. (1980). The person-centered approach to therapy. Presentation at the Scottish Association for Therapy, Glasgow, United Kingdom. Retrieved from http://www.elementsuk.com/libraryofarticles/thepcato.pdf

Mearns, D., & Cooper, M. (2005). *Working at relational depth in counselling and psychotherapy*. London, United Kingdom: Sage.

Merry, T. (2001). Congruence and the supervision of client centred therapists. In G. Wyatt (Ed.), Congruence *(Rogers' therapeutic conditions: Evolution, theory and practice)* (Vol. 1, pp. 174–183). Ross-on-Wye, United Kingdom: PCCS Books.

Miller, S. D., Duncan, B. L., & Hubble, M. A. (1997). *Escape from Babel: Toward a unifying language for psychotherapy practice*. New York, NY: Norton.

Moon, K. A., Witty, M., Grant, B., & Rice, B. (2011). *Practicing client-centered therapy: Selected writings of Barbara Temaner Brodley*. Ross-on-Wye, United Kingdom: PCCS Books.

Mostchnig-Pitrik, R., Lux, M., & Cornelius-White, J. H. D. (2013). The person-centered approach: An emergent paradigm. In J. Cornelius-White, R. Motschnig-Pitrik, & M. Lux (Eds.), *Interdisciplinary applications of the person-centered approach* (pp. 235–251). New York, NY: Springer.

Multicultural Counseling and Social Justice Competencies. (2013). Social justice. Retrieved from http://toporek.org/social_justice.html

Norcross, J. C. (2002). *Psychotherapy relationships that work: Therapist contributions and responsiveness to patients*. New York, NY: Oxford University Press.

Norcross, J. (2005). The psychotherapist's own psychotherapy: Educating and developing psychologists. *American Psychologist, 60*(8), 840–850.

O'Hara, M. (2003). Cultivating consciousness: Carl R. Rogers's person-centered group process as transformative androgyny. *Journal of Transformative Education, 1*(1), 64–79. Retrieved from http://insightu.net/content/library/journals/jtevol01no01january200364-79.pdf

Patterson, C. H. (1984). Empathy, warmth, and genuineness in psychotherapy: A review of reviews. *Psychotherapy, 21*, 431–438.

Patterson, C. H. (1996). Multicultural counseling: From diversity to universality. *Journal of Counseling and Development, 74*, 227–231. Also published in (2000) *Understanding psychotherapy: Fifty years of client-centered theory and practice*. Ross-on-Wye, United Kingdom: PCCS Books. Retrieved from https://www.tiaa-cref.org/public/support/forms/topics/qualified_dom_rel_order.html

PCAYorks. (2014). Carl Rogers' transcripts. Retrieved from http://pcayorks.blogspot.com/2010/02/carl-rogers-transcripts.html

PCE. (2010). Empowerment: The politics of the helping relationship. 9th World Conference for Person-Centered and Experiential Psychotherapy and Counseling. Retrieved from http://www.docstoc.com/docs/85250764/PCE-2010-Empowerment-The-politics-of-the-helping-relationship or www.pce-org

Pope, M. (2002). Counseling individuals from the lesbian and gay cultures. In J. Trusty, E. J. Looby, & D. S. Sandhu (Eds.), *Multicultural counseling: Context, theory and practice, and competence* (pp. 201–218). Huntington, NY: Nova Science.

Proctor, G. (2002). *The dynamics of power in counseling and psychotherapy: Ethics, politics, and practice.* Ross-on-Wye, United Kingdom: PCCS Books.

Proctor, G., Cooper, M., Sanders, P., & Malcolm, B. (Eds.). (2006). *Politicizing the person-centered approach: An agenda for social change.* Ross-on-Wye, United Kingdom: PCCS Books.

Prouty, G. (1994). *Theoretical evolutions in person-centered/experiential therapy: Applications to schizophrenic and retarded psychoses.* New York, NY: Praeger.

Rahula, W. (2007). *What the Buddha taught.* New York, NY: Grove.

Rice, B. (2013). Foundational oppression: Families and schools. In J. H. D. Cornelius-White, R. Mostnig-Pitrik, & M. Lux (Eds.), *Interdisciplinary applications of the person-centered approach* (pp. 141–144). New York, NY: Springer.

Rice, L. N. (1974). The evocative function of the therapist. In D. Wexler & L. Rice (Eds.), *Innovations in client-centered therapy* (pp. 289–311). New York, NY: Wiley.

Richardson, T. Q., & Jacob, E. J. (2002). Contemporary issues in multicultural counseling: Training competent counselors. In J. Trusty, E. J. Looby, & D. S. Sandhu (Eds.), *Multicultural counseling: Context, theory and practice, and competence* (pp. 31–45). Huntington, NY: Nova Science.

Rogers, C. R. (1951). *Client-centered therapy: Its current practice, implications, and theory.* London, United Kingdom: Constable.

Rogers, C. R. (1957). The necessary and sufficient conditions of therapeutic personality change. *Journal of Consulting Psychology, 21*(2), 95–103.

Rogers, C. R. (1959). A theory of therapy, personality and interpersonal relationships as developed in the client-centered framework. In S. Koch (Ed.), *Psychology: A study of a science. Vol. 3: Formulations of the person and the social context.* New York, NY: McGraw-Hill.

Rogers, C. R. (1961). *On becoming a person: A therapist's view of psychotherapy.* London, United Kingdom: Constable.

Rogers, C. R. (1977). *On personal power: Inner strength and its revolutionary impact.* Boston, MA: McGraw-Hill.

Rogers, C. R. (1980). *A way of being.* Boston, MA: Houghton Mifflin.

Rogers, C. R. (1983). *Freedom to learn for the 80s.* Columbus, OH: Charles Merrill.

Rogers, C. R., & Cornelius-White, J. H. D. (2013). *Carl Rogers: The China diary.* Charleston, SC: Create Space.

Rogers, C. R., Lyon, H. C., & Tausch, R. (2013). *On becoming an effective teacher: Person-centered teaching, psychology, philosophy, and dialogues with Carl R. Rogers and Harold Lyon.* New York, NY: Routledge.

Rosenzweig, S. (1936). Some implicit common factors in diverse methods of psychotherapy. *American Journal of Orthopsychiatry, 6*, 412–415. doi: 10.1111/j.1939-0025.1936.tb05248.x

Sanders, P. (2004). *The tribes of the person-centered nation: An introduction to the schools of therapy related to the person-centered approach.* Ross-on-Wye, United Kingdom: PCCS Books.

Schmid, P. F. (2001). Authenticity: The person as his or her own author. Dialogical and ethical perspectives on therapy as an encounter relationship. In G. Wyatt (Ed.), *Congruence (Rogers' therapeutic conditions: Evolution, theory and practice)* (Vol. I, pp. 217–232). Ross-on-Wye, United Kingdom: PCCS Books.

Schmid, P. (2013). A practice of social ethics: Anthropological, epistemological, and ethical foundations of the person-centered approach. In J. Cornelius-White, R. Motschnig-Pitrik, & M. Lux (Eds.), *Interdisciplinary handbook of the person-centered approach: Research and theory* (pp. 353–368). New York, NY: Springer.

Smith, M. L., & Glass, G. V. (1977). Meta-analysis of psychotherapy outcome studies. *American Psychologist, 32*, 752–760.

Smith, T. B., & Richards, P. S. (2002). Multicultural counseling in spiritual and religious contexts. In J. Trusty, E. J. Looby, & D. S. Sandhu (Eds.), *Multicultural counseling: Context, theory and practice, and competence* (pp. 105–128). Huntington, NY: Nova Science.

Sommerbeck, L. (2006). Udenfor Terapeutisk Rækkevidde? Introduktion til Præ-Terapi [Beyond psychotherapeutic reach? An introduction to pre-therapy]. *Psykolog Nyt, 60*(8), 12–20.

Stubbs, J. P., & Bozarth, J. D. (1994). The dodo bird revisited: A qualitative study of psychotherapy efficacy research. *Journal of Applied and Preventive Psychology, 3*(2), 109–120.

Sue, D. W. (2001). Multidimensional facets of cultural competence. *Counseling Psychologist, 29*, 790–821. doi: 10.1177/0011000001296002

Sue, D. W., Arredondo, P., & McDavis, R. J. (1992). Multicultural counseling competencies and standards: A call to the profession. *Journal of Counseling & Development, 70*(4), 477–486.

Sue, D. W., Capodilupo, C. M., Torino, G. C., Bucceri, J. M., Holder, A. M. B., Nadal, K. L., & Esquilin, M. (2007). Racial microaggressions in everyday life: Implications for clinical practice. *American Psychologist, 62*(4), 271–286.

Sue, D. W., & Sue, D. (2012). *Counseling the culturally diverse: Theory and practice.* New York, NY: Wiley.

Toporek, R. L., Lewis, J. A., & Crethar, H. C. (2009). Promoting systemic change through the ACA Advocacy Competencies. *Journal of Counseling & Development, 87*, 260–268.

Wampold, B. E. (2001). *The great psychotherapy debate: Models, methods, and findings.* Mahwah, NJ: Erlbaum.

Wampold, B. E., Mondin, G. W., Moody, M., Stich, F., Benson, K., & Ahn, H. (1997). A meta-analysis of outcome studies comparing bona fide psychotherapies: Empirically, "All must have prizes." *Psychological Bulletin, 122*, 203–215.

World Association for Person-Centered and Experiential Psychotherapy and Counseling. (2000). Statutes and bylaws. Retrieved from http://www.pce-world.org/about-us/statutes-and-bylaws.html

# Index

Figures and tables are indicated by f or t following the page number.

Action without action, 48
Actualizing tendency, 12, 18–19, 71
Adaptability to differences, 50–51
ADDRESSING model, xii
Advocacy, 45–46, 45t
American Counseling Association, 39
Anna O. (Pappenheim, Bertha), x
Attachment, therapeutic, 5
Attrition of clients, 18
Authenticity, 3
Authority, experience as basis for, 23

Baldwin, M, 8
Barrett-Lennard, G., 29
"Being with" clients, xix, xxii, 12, 68–69
Benign neutrality, 5
Biases. *See* Multiculturalism
Boundaries, 12
Brodley, B. T., 19
Bucceri, J. M., 43
Buddhism, 49

Capodilupo, C. M., 43
*Carl Rogers Reader* (Kirschenbaum & Henderson), xviii
Case illustration
  first session, 54–56
  second session, 56–58
  third session, 58
  fourth session, 58–60
  fifth session, 60–61
  sixth session, 61–63

  seventh session, 63–64
  missed sessions, 64
  client, 53
  Outcome Rating Scale, 65f
  Session Rating Scale, 65f
  therapist, 53–54
Casey. *See* Case illustration
Christianity, 48–49
Circles of contact, 33–34
Clarification process, 6–8. *See also* Feedback loops
Classical client-centered therapy, 24–27
Client-centered approaches. *See* Person-centered approaches (PCA)
Client-directed approaches, 17–18
Client perceptions. *See* Perception
Clients
  attrition of, 18
  feedback from, 6–8, 16–18, 17t, 68–69
  necessary and sufficient conditions for, 13–17, 13t, 16–17t
  role in therapeutic outcomes, 11–12
  therapeutic change percentage attributed to, x
  therapists as, 11, 20–21
Common factors hypothesis
  attrition of clients, 18
  classical client-centered therapy and, 25–26
  experiential emotion-focused approach, 32
  focusing and, 27–28

81

intersubjectivity and, 29–30
overview, ix, x, xii, xiii
specificity myth and, 39–40
Compassionate confrontation, 68
Congruence
  as cultural, 40–41
  defined, 3
  integration of, 8–9, 8–9f
  overview, 2–4, 3t
Contact, 13–14, 13t, 68
Contact responses, 14
Cooper, M., 29
Core conditions, 2, 8–9, 8–9f. *See also* Congruence; Empathic understanding; Unconditional positive regard (UPR)
Cornelius-White, J. H. D., 2–3
Counselors. *See* Therapists
Courage, 12, 68
Cultural competence. *See* Multiculturalism

Deena. *See* Case illustration
Differences, adaptability to, 50–51
Dione, 26–27
Discrimination, 40. *See also* Multiculturalism
Diversity. *See* Multiculturalism
Dodo bird verdict, 18
Dropout rates, 18
Duncan, B. L., xii, xx, 18

Effortless doing, 48
Emotion-focused approach, 30–33
Emotions, 31, 68
Empathic understanding
  client perceptions and, 69f
  cultural issues, 42–43, 43t
  importance of, xxi
  love and, 67
  overview, 5–9, 6t, 8–9f
Empowerment, 45–46, 45t, 70–71
Empty chair technique, 32–33
Esquilin, M., 43
Ethics, 24–25, 47
Evolution of person-centered approaches
  classical client-centered therapy, 24–27

experiential emotion-focused approach, 30–33
focusing, 27–28
futurism, 36–37
interdisciplinary practice, 33–36, 34–35f
intersubjectivity, 28–30
overview, 23–24
Experiential emotion-focused approach, 30–33
Experiential psychotherapy, 27
Experiential researchers, xx

Feedback loops, 6–8, 16–18, 17t, 68–69
Focusing, 27–28
*Focusing* (Gendlin), 28
Freud, Sigmund, ix, x–xii

Game theory, 34
Gendlin, Eugene, 27–28
Genuineness, 67
Goals, 49–50
Grace, 48–49
Grit, 19

Hate crimes, 40
Henderson, V., xviii
*The Heroic Client* (Duncan & Miller), xii
Holder, A. M. B., 43
Holocaust, xi
Houston, 44
Hubble, M. A., 18

Incongruence, 14–15, 16t, 68, 69f
*Incurvatus in se*, 49
Integrated, 3
Interdisciplinary practice, 33–36, 34–35f
Intersubjectivity, 28–30
Intervention selection, 39–40

Kirschenbaum, H., xviii

Lao-tzu, 19, 48
Learner-centered education, xviii
A List of Privilege Lists, 43
Los Angeles, 44

Love, 67
Lux, M., 33–34

Markers, 31
McIntosh, P., 42–43
Mearns, D., 23, 29
Merry, T., 41
Meta-analysis, xiii
Microaggressions, 43–45, 44t
Miller, S. D., xii, xx, 18
Mindfulness, 27–28, 49–50
Mirror neurons, xxi, 35
Morality, 24–25
Multiculturalism
   adaptability to differences, 50–51
   backdrop for, xi
   congruence and, 40–41
   defined, 39
   empathic understanding and, 42–43, 43t
   history of, 46–47
   importance of, xxiii
   overview, 39–40
   religion and, 47–50
   social justice, 45–46, 45t
   stereotypes, 41–42, 42t
   unconditional positive regard and, 43–45, 44t
   understanding of clients and, xiii–xix

Nadal, K. L., 43
Necessary and sufficient conditions
   in classical client-centered therapy, 24
   for clients, 13–17, 13t, 16–17t
   for relationships, 2–9, 3t, 5–6t, 8–9f
"The Necessary and Sufficient Conditions" (Rogers), xix
Neurons, xxiii, 35
Neuroscience, xxiii, 33–35
Neutrality, 5
Nonattachment, 5, 49–50
Nondirective attitude, 12, 19–20, 24–25

O'Hara, M., 36–37
Outcome-informed approaches, 17–18
Outcome Rating Scale (ORS), 18, 54, 70

Pappenheim, Bertha, x
Perception
   as core condition, 16–17, 17t
   empathic understanding and, 69f
   importance of, xxiv, 16, 68
Person-centered approaches (PCA)
   futurism and, 36–37
   history of, 46–47
   levels of, 34–35, 35f
   as love, 67
   minority participants and, 47
   scope of, 33, 34f
   therapists choosing for self, 20–21
   *See also* Case illustration; Evolution of person-centered approaches
Prejudices. *See* Multiculturalism
Presence, 8–9
Pretherapy, 13
Primary adaptive emotions, 31
Primary maladaptive emotions, 31
Privilege, 42–43, 43t
Privilege Lists, 43
Process directive, 31

Racial issues, xiii. *See also* Multiculturalism
Raskin, N., 25
Rational self-interest, 34
Relationships
   appreciation of, 1
   in classical client-centered therapy, 25–26
   importance of, xxi
   necessary and sufficient conditions for, 2–9, 3t, 5–6t, 8–9f
   therapeutic change percentage attributed to, x
Religion, 47–50
RESPECTFUL framework, xi
Retention of clients, 18
Rogerian, 23
Rogers, Carl
   biography, xix–xx, 46–47
   on congruence, 40
   Dione and, 26–27
   on experience, 23
   as futurist, 36–37
   influence of, xx
   multiculturalism and, xxi
   on nondirective attitude, 19, 25
   on presence, 8

Schmid, P. F., 29, 40–41
Secondary emotions, 31
Session Rating Scale (SRS), 17–18, 54, 70
Social action, 45
Social justice, 45–46, 45t, 70–71
Specificity myth, 39–40
Spiritualism, 47–50
Stereotypes, 41–42, 42t
Sue, D. W., 43

*Tao Te Ching* (Lao-tzu), 19, 48
Teachingtolerance.org, 41
Theories, xi–xii
Therapeutic attachment, 5
Therapeutic nonattachment, 5
Therapeutic outcomes, 1–2, 11–12
Therapeutic relationship. *See* Relationships
Therapists
  actualizing tendency, 71
  "being with" client, xix, xxii, 12, 68–69
  in classical client-centered therapy, 25–26
  as clients, 11, 20–21
  effectiveness of, xiii
  goals for clients, 49–50
  perception of client not appreciated by, 16
  therapeutic outcomes and, 1–2, 12
Torino, G. C., 43
Transparency, 3

Unconditional positive regard (UPR)
  cultural issues, 43–45, 44t
  importance of, 67, 69–70, 69f
  integration of, 8–9, 8–9f
  overview, 4–5, 5t
Understanding, 70. *See also* Empathic understanding; Feedback loops

Vulnerability, 15

Walking in someone else's shoes. *See* Empathic understanding
Wampold, B. E., 18
Wei-wu-wei, 48
White privilege, 42–43
World Association of Person-Centered and Experiential Psychotherapy and Counseling, 23

# About the Author

**Jeffrey H. D. Cornelius-White**, PsyD, LPC, is Professor and Counseling Programs Coordinator in the Department of Counseling, Leadership and Special Education at Missouri State University in Springfield, and Adjunct Assistant Professor and doctoral faculty for the Cooperative EdD Program in Educational Leadership and Policy Analysis at the University of Missouri–Columbia. He served as Chair of the Board of the World Association for Person-Centered and Experiential Psychotherapy and Counseling, Coeditor of *The Person-Centered Journal*, Content Editor of the *Journal of Border Educational Research*, and has edited and authored several other books, including *Learner-Centered Instruction* with SAGE (2010). His 100 publications have been mostly concerned with person-centered and social justice issues in education, psychology, and counseling. Jef is the father of two children, Avery and Evan. He enjoys biking, swimming, reading, and playing with his kids.

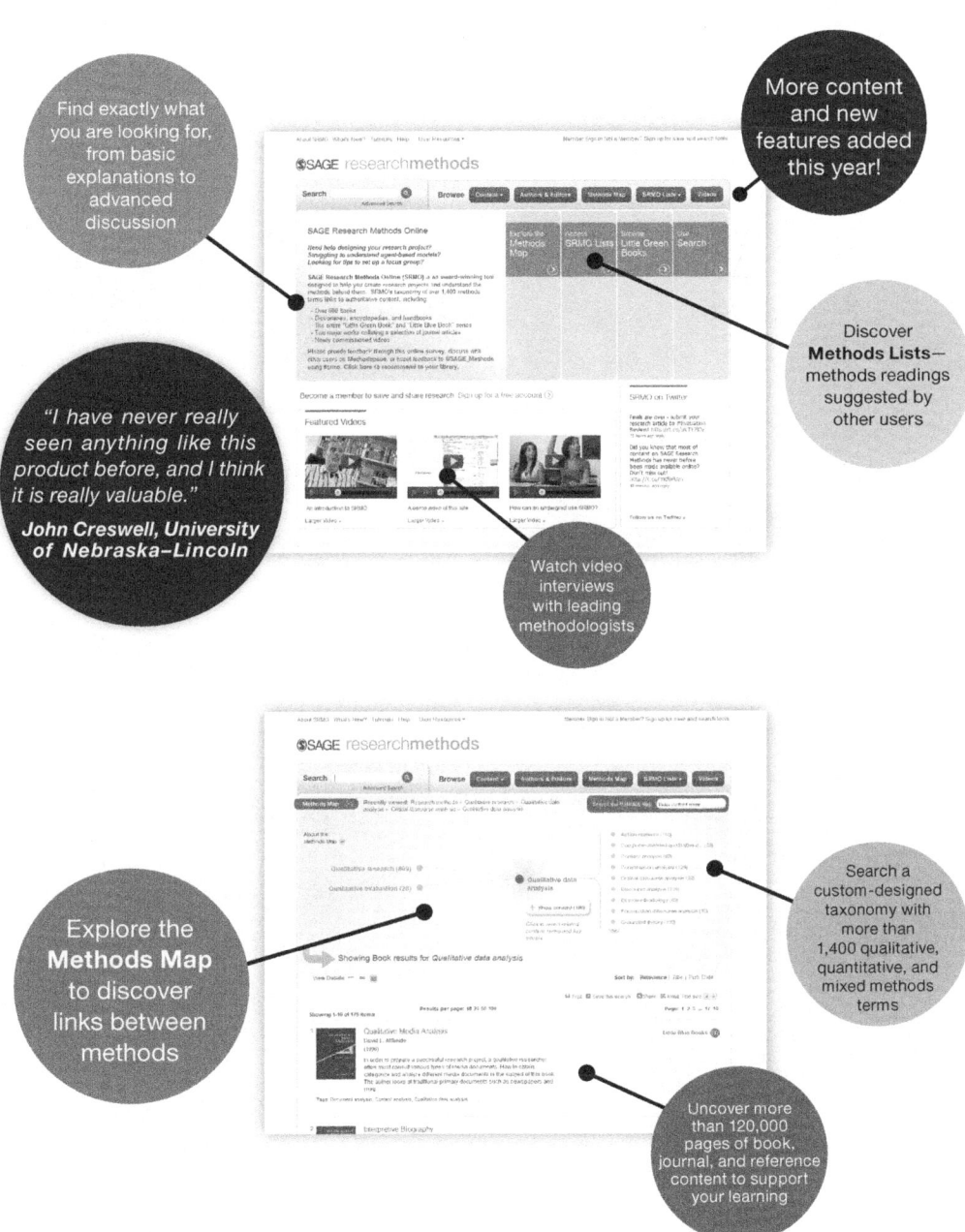

Made in the USA
Monee, IL
29 August 2023

41837885R00066